Off The Grid Prepper's Water Survival Plan

7-Steps to Find, Purify, Filter and Safely Store for Self-Sufficiency and Cost-Savings

By

Andrew Raines

Copyright 2024 - **All rights reserved.** The content contained within this book may not be reproduced, duplicated or transmitted without direct written permission from the author or the publisher. Under no circumstances will any blame or legal responsibility be held against the publisher, or author, for any damages, reparation, or monetary loss due to the information contained within this book, either directly or indirectly.

Legal Notice:

This book is copyright protected. It is only for personal use. You cannot amend, distribute, sell, use, quote or paraphrase any part, or the content within this book, without the consent of the author or publisher.

Disclaimer Notice:

Please note the information contained within this document is for educational and entertainment purposes only. All effort has been executed to present accurate, up to date, reliable, complete information. No warranties of any kind are declared or implied. Readers acknowledge that the author is not engaged in the rendering of legal, financial, medical or professional advice. The content within this book has been derived from various sources. Please consult a licensed professional before attempting any techniques outlined in this book.

By reading this document, the reader agrees that under no circumstances is the author responsible for any losses, direct or indirect, that are incurred as a result of the use of the information contained within this document, including, but not limited to, errors, omissions, or inaccuracies.

Contents

Introduction	1
1. Water and Self-Sufficiency	5
2. Find and Collect	21
3. Water Purification	47
4. Long Term Storage	71
5. Prevent Waterborne Diseases	99
6. Innovative Water Conservation Practices	113
7. 7-Step Water Survival Plan	131
Bonus Chapter 1- Preparing For Emergencies	147
Bonus Chapter 2 - Water-Saving Tools	163
Conclusion	173
References	175

INTRODUCTION

As we go about our daily lives, we often encounter unexpected challenges and moments of calmness. During these unpredictable twists of fate, the importance of preparedness becomes evident. Benjamin Franklin once wisely said, "By failing to prepare, you are preparing to fail" (Mealeatey, 2021). This statement is particularly relevant when it comes to emergencies, where our ability to survive is closely linked to our level of preparedness, especially when it comes to securing a resource as essential as water.

Imagine a world where water scarcity is not a distant possibility but a harsh reality faced by millions, if not billions, of people daily. Shocking statistics predict that the number of urban populations grappling with water scarcity will double from 930 million in 2016 to a staggering 1.7–2.4 billion by 2050, according to UNESCO (2023). This isn't just a statistic to ignore; it's a warning that the life-sustaining liquid we take for granted may soon become a rare and valuable commodity.

This book delves into the critical link between preparation and survival in water-related emergencies. It offers guidance on navigating challenging situations and provides valuable information to help readers stay safe. The book emphasizes the importance of understanding the gravity of Franklin's words as the seas of uncertainty continue to rise. As readers turn the pages, they will discover the challenges of water scarcity and the wide-ranging consequences that ripple through societies unprepared for the impending storm.

Read and Reap the Benefits

I understand that making the most of the available water resources can be challenging, especially when dealing with emergencies and facing uncertainties. I realize the gravity of your concerns, the urgent need to guarantee the safety and well-being of yourself and your loved ones, and the importance of confidently navigating the intricacies of off-grid water survival.

Off The Grid - Prepper's Water Survival Plan is a comprehensive guide that can be your trustworthy companion on your journey towards achieving water self-sufficiency. Water-related challenges are becoming more frequent and severe, making it essential to have practical knowledge and solutions at hand. This book provides just that, with tangible strategies to help you navigate through difficult times.

Why should you explore with me? Consider the benefits of these pages:

- *Discover unlikely sources:* Unravel the secrets of finding and purifying water from the most unlikely sources. In times of scarcity, knowing where to look and how to harness nature's hidden reservoirs can be the key to sustaining yourself and those around you.

- *Natural purification techniques:* Reduce your reliance on modern technology by learning natural ways to purify water. These time-tested methods will ensure that even without high-tech devices, you can transform questionable water into a safe and life-sustaining resource.

- *Essential long-term storage skills:* Learn to store water safely and effectively. Choose the best containers for long-term storage to secure your water supply.

As a retired civil engineer with years of experience in water resource management, I understand the challenges you may face and have used my expertise and knowledge to create a comprehensive guide that will be helpful to you. This guide contains practical information that you can apply to real-world scenarios.

The *7-Step Water Survival Plan* that accompanies this book is not just a guide but a companion that will stand with you on your journey to water survival. The information within these pages is not just theoretical but is instead based on practical experience.

We will discuss a future where water is scarce and explore the complex relationship between preparedness and the impending water crisis. Each chapter will provide you with valuable knowledge that will help you navigate through uncertain times. Together, let's pave the way for a future where we can guarantee that failure is not an option but an opportunity for growth and resilience.

Your water survival empowerment journey begins now.

***

***

1

WATER AND SELF-SUFFICIENCY

Water has always been crucial to life on Earth. As Jacques Yves Cousteau once said, "We forget that the water cycle and the life cycle are one" (Oberle, 2020). This illustrates how our lives are intertwined with water in a relationship that has existed since the beginning of time.

Imagine a world without water—no rivers, lakes, or oceans. It would be a bleak and lifeless place devoid of the vitality that water brings. But water isn't just a background element; it's the very essence of life.

At first glance, water may seem commonplace, a resource so abundant that its significance is easily overlooked. Yet, beneath this veneer of familiarity lies a profound truth—water is the elixir that sustains all living beings. It is the source of life, a universal solvent that nurtures and nourishes, connecting every living organism in a delicate web of interdependence.

Our daily routines, such as drinking water and tending to crops, are simple acknowledgments of water's vital role in our survival. However, it's easy to overlook the immense impact that water has on our lives and to take for granted that our existence is closely connected to the ebb and flow of this essential life force.

Water is essential for our bodies and health. It is also crucial to the natural water cycle. When we drink water, we replenish our bodies and acknowledge the continuous cycle of life and renewal. It's fascinating to think about how water droplets travel through the complex pathways of nature, transitioning from vaporous clouds to liquid streams and back again. This cyclical journey of water reflects the constant cycle of life and renewal.

Water's importance goes beyond personal needs. In many agricultural areas, it turns dry lands into fertile fields, nourishing crops that sustain entire communities. In the wild, it is the life source of ecosystems, supporting diverse flora and fauna and creating habitats that are full of vitality.

We must recognize that water is not a mere commodity, a resource to be exploited without consequence. It is a fundamental element, an irreplaceable contributor to the delicate balance that allows life to flourish on our blue planet. To safeguard our future, we must appreciate the interconnectedness of water and life cycles.

Remember that water is not just something we use—it is essential to life. As we go about our lives, we must take responsibility for preserving and cherishing this vital resource that sustains all living things. Remember this as we engage in discussion, and remember that our actions can make a difference.

Water Cycle: The Natural Movement of Water

Understanding the water cycle is crucial to utilizing our water resources effectively. The water cycle comprises several stages, including evaporation, condensation, and precipitation. By understanding this process, we can make informed decisions about the sustainable management of water. This means using water in a way that does not harm the environment and can be maintained over time. Understanding the water cycle is the first step in ensuring we use water as efficiently as possible.

The water cycle begins in the oceans, lakes, and rivers:

Evaporation

The sun's radiant heat warms the surface of the water bodies, causing water molecules to gain enough kinetic energy to break free from their liquid bonds, transforming into vapor.

Condensation

Water vapor rises into the atmosphere, where cooler temperatures exist. At this high altitude, it cools down and turns into tiny water droplets through condensation. The droplets then gather together to form clouds.

Precipitation

When clouds gather and become fully loaded with water, the tiny water droplets within them join together to form larger droplets. These drops then fall from the sky due to gravity and form precipitation, which can take different forms, such as a light drizzle that nourishes the earth or a heavy downpour that shapes the landscape.

Infiltration

When precipitation, such as rain or snow, falls onto the ground, it can be absorbed by the soil through infiltration. This water that infiltrates the soil is essential for plant growth, replenishing the groundwater reserves, and feeding into the complex network of aquifers that support ecosystems.

Surface Runoff

Water that falls on the ground can either seep into the soil or flow over the land surface through creeks and streams. This water flow is known as surface runoff. As it travels towards larger bodies of water, surface runoff gains momentum and shapes the landscape. It plays a vital role in nourishing the intricate web of life as well as carving valleys and shaping the overall topography.

Transpiration

Plants absorb water through their roots and then release the excess moisture through tiny openings in their leaves in a process called transpiration. This process adds to the water vapor in the atmosphere, completing a cycle that connects plants to the larger water cycle.

Sublimation and Freezing

At extremely low temperatures, water can undergo sublimation, transitioning directly from a solid state (ice or snow) to a gas without passing through the liquid state. Conversely, freezing occurs when water molecules lose energy and transform into ice. This process contributes to the icy landscapes seen in polar regions.

The water cycle is a complex process demonstrating the interconnectedness of Earth's systems. This ongoing cycle supports life, shapes landscapes, and showcases our planet's delicate balance. As we appreciate the beauty of this intricate dance, we should also acknowledge the profound truth that the water and life cycles are one and the same, forever linked in the remarkable story of our world.

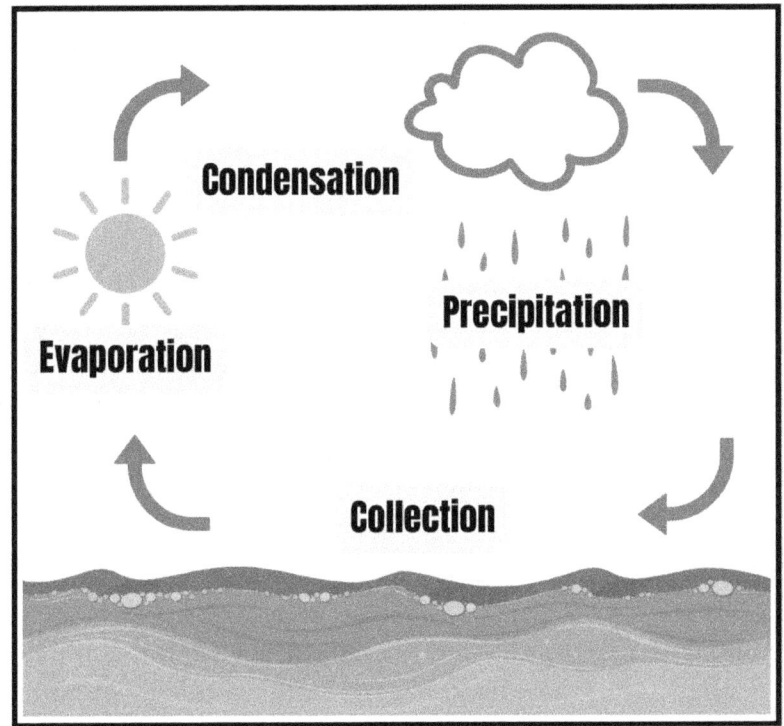
Water Cycle

Global Water Crisis

As you delve into the subject of water survival, it is crucial to understand the current global water crisis. The present challenges present a grim picture, highlighting the precarious state of our water resources. People worldwide are struggling with increasing water scarcity caused by a rise in population, rapid urbanization, and unsustainable agricultural practices. On top of these challenges is the constant threat of climate change, which is a significant disruptor to the fragile balance of water availability.

Climate change is an unstoppable force altering our planet's dynamics. It worsens water crises by causing more severe and frequent droughts, floods, and unpredictable precipitation patterns.

Higher temperatures cause more evaporation, depleting surface water sources. Extreme weather events—becoming more intense due to climate change—directly threaten water quality as they can contaminate water supplies with pollutants. This puts the health of entire communities at risk.

According to reputable sources like the Council on Foreign Relations, the global water crisis is not just about scarcity (Klobucista & Robinson, 2023). Water quality is also affected, with industrial discharges, agricultural runoff, and inadequate sanitation systems polluting freshwater sources. This harms ecosystems and affects the health and well-being of human populations that rely on these contaminated water supplies

Water in Daily Life

Although it may seem obvious, we must remember the significance of water in our daily lives. It plays a crucial role in sustaining ecosystems and, inevitably, human existence. Water is the lifeblood of the ecosystem, maintaining a delicate balance that supports the diverse web of life on Earth. In its natural cycle, water nourishes flora and fauna, enabling the flourishing of ecosystems from lush rainforests to arid deserts (Juste, 2023).

Water's importance goes beyond its ecological impact, as it permeates every aspect of our daily activities. Lack of access to water or scarcity can profoundly impact our lives, disrupting the fabric of our routines and well-being. For instance, the simple act of staying hydrated is essential for human survival. Without an adequate and accessible water supply, our health is jeopardized, leading to dehydration, weakened immune systems, and a host of associated health issues.

Moreover, agriculture, a cornerstone of human sustenance, is intricately tied to water availability. Crops, the backbone of our food supply, rely on consistent and sufficient water to thrive. In regions where water is scarce, agricultural productivity dwindles, leading to food shortages, price hikes, and economic instability. The implications of water scarcity ripple through global supply chains, affecting communities far beyond the immediate locus of water stress.

Water plays a pivotal role in production processes in the industrial and manufacturing sectors. Water scarcity disrupts these operations, hampering economic activities and industry growth. Access to clean water is also integral to sanitation and hygiene, crucial components of public health. Without proper water resources, the risk of waterborne diseases escalates, affecting communities with dire consequences

Furthermore, ecosystems are intricately interconnected, and water scarcity can lead to habitat degradation and loss of biodiversity. Aquatic ecosystems, in particular, suffer when water availability is compromised, leading to the decline of fish populations and disrupting the delicate balance of marine and freshwater ecosystems.

Water scarcity or unavailability can negatively affect our physical health, economic stability, and the environment. It vividly reminds us of the interconnectedness of all life on Earth and water's significant role in our daily lives.

Role of Water in Overall Well-Being

Though water is mostly perceived as a beverage to quench our thirst, let's not forget that getting hydrated is a cornerstone of overall health and well-being. Adequate water intake is paramount for optimal physiological functioning, and its importance cannot be overstated. Most illnesses, ranging from mild discomfort to severe conditions, could be mitigated or even avoided if individuals maintained an ideal water intake.

Hydration is fundamental to the proper functioning of various bodily systems. Water is the primary component of cells, tissues, and organs, facilitating essential biological processes. It aids digestion, enabling the breakdown of nutrients and the absorption of vital elements in the gastrointestinal tract. Proper hydration also supports kidney function by assisting in the filtration of waste products, ensuring the elimination of toxins from the body (Zahmak, 2013).

Maintaining an optimal water balance is crucial for cardiovascular health. Water contributes to the composition of blood, ensuring its viscosity and flow. Dehydration can strain the heart, leading to an increased heart rate and elevated blood pressure. By staying adequately hydrated, individuals can mitigate the risk of cardiovascular issues and support the overall efficiency of their circulatory system.

One of the most apparent benefits of proper hydration is its impact on cognitive function. The brain is highly sensitive to changes in hydration levels, and even mild dehydration can impair concentration, alertness, and short-term memory. Adequate water intake promotes mental clarity, focus, and overall cognitive performance.

Furthermore, water plays a crucial role in regulating body temperature. Through the process of sweating, the body dissipates heat, preventing overheating during physical activity or in warm environments. Dehydration impairs this cooling mechanism, increasing the risk of heat-related illnesses such as heat exhaustion and heatstroke.

In preventative healthcare, maintaining ideal water intake can be a powerful ally. Dehydration weakens the immune system, making individuals more susceptible to infections and illnesses. Adequate hydration supports immune function, aiding the body in its defense against pathogens and reducing the likelihood of falling ill (Garcia-Garcia, 2022).

Chronic conditions, such as kidney stones and urinary tract infections, are often linked to insufficient water intake. Water serves as a natural cleanser for the urinary tract, flushing out potentially harmful substances and preventing the formation of crystals and stones.

Water is not just a drink but a crucial component of our health and well-being. Proper hydration significantly impacts our body's functions, and we can take proactive measures to ensure that our body stays hydrated. Maintaining an adequate and consistent water intake can prevent or alleviate most illnesses, from minor to severe. A simple yet powerful way to improve our overall health is to commit to staying well-hydrated, which can lead to vitality and longevity.

Self-Sufficiency in Survival

You hopped into this journey with survival in mind. Therefore, we need to discuss self-sufficiency, which is the key to resilience and adaptability. Self-sufficiency involves the ability to meet your needs independently, without relying heavily on external resources or assistance. Picture it as a survival toolkit—a set of skills and mindset that empowers you to navigate challenges effectively. Self-sufficiency is not just a practical skill; it's a philosophy that places emphasis on autonomy and preparedness.

The importance of self-sufficiency becomes apparent when facing unforeseen circumstances, be it in the wilderness or in the midst of a crisis. By cultivating self-reliance, you become less dependent on external systems, ensuring your ability to endure and overcome challenges. In survival scenarios, where resources may be scarce or infrastructure compromised, the self-sufficient individual stands better equipped to adapt and persevere.

In the specific realm of water survival, imagine a scenario where water resources are scarce, contaminated, or disrupted. This situation is not far-fetched in the current global water crisis. In such circumstances, being able to source, purify, and manage water independently becomes crucial.

Regarding the broader discussion on the water crisis, self-sufficiency becomes a strategic response to water scarcity and compromised water quality challenges. By mastering the skills of sourcing and purifying water, one ensures one's own survival and contributes to a more sustainable use of water resources. Therefore, self-sufficiency in water survival aligns with the broader imperative of responsible water management. This helps mitigate the water crisis's impact on both individual and collective levels.

It is becoming increasingly clear that embracing self-sufficiency is not just a survival tactic but a mindset that empowers individuals to thrive in challenging situations. This is especially true in water survival and the global water crisis, where self-sufficiency is integral to resilience and weathering the storm. By cultivating self-sufficiency, you can navigate challenges with greater confidence and agility and help contribute to securing water for the future. So take control of your survival journey, embrace self-sufficiency, and navigate the challenges confidently and quickly.

Traditional Versus Off-Grid Water Systems

To achieve our goal of water independence, it is crucial to understand the distinction between a traditional water supply system and an off-grid system. Most properties receive drinkable water from publicly owned utilities through a traditional system. Conversely, off-grid systems are small private networks catering to individual properties or small communities.

Traditional Water Supply System

Traditionally, a water supply system is a network that provides clean water to homes and communities. The system includes pipes, water tanks, pumps, and treatment facilities that ensure continuous and safe delivery of drinkable water for our daily needs.

This comprehensive infrastructure is responsible for sourcing, treating, and distributing water, making it an essential part of modern living.

In a typical water system:

- *Sourcing water:* Water is typically sourced from rivers, lakes, or underground aquifers.

- *Treatment:* The water undergoes treatment to remove impurities, contaminants, and potential health hazards.

- *Distribution:* Through a network of pipes, pumps, and water tanks, water is distributed to homes, schools, businesses, and beyond.

- *Wastewater management:* After use, wastewater is collected, treated, and returned to the environment, or in some cases, reused.

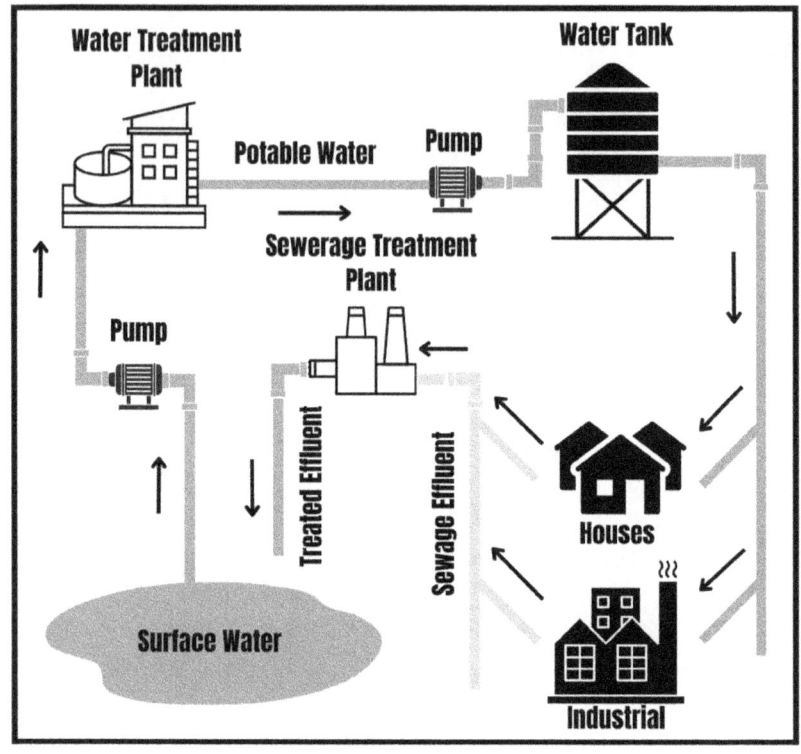

Traditional Water Supply System

Off-Grid Water Supply System

Now, let's shift our focus to the concept of an off-grid water supply system. Unlike the conventional setups integrated into municipal infrastructure, an off-grid system operates independently, often in remote or rural areas. The off-grid approach champions self-sufficiency, allowing individuals or communities to manage their water needs without reliance on external utilities.

Here are the key points that make an off-grid water system different:

- *Independent sourcing:* Off-grid systems often rely on local sources like wells, springs, or rainwater harvesting. This autonomy in sourcing water distinguishes them from centralized systems.

- *On-site treatment:* Rather than relying on distant treatment facilities, off-grid systems incorporate on-site treatment methods. This can include filtration, purification, and disinfection technologies tailored to the specific water source.

- *Local distribution:* The distribution network in off-grid systems is more localized, with water being directly supplied to individual homes or smaller clusters. This decentralized approach minimizes the need for extensive piping infrastructure.

- *Sustainability focus:* Off-grid systems often prioritize sustainability, incorporating eco-friendly practices such as rainwater harvesting, graywater reuse, and energy-efficient technologies.

- *Challenges and solutions:* Off-grid systems may face challenges like maintenance and expertise while offering independence. However, technological advancements and community-driven solutions address these issues, making off-grid water systems increasingly viable.

In summary, an off-grid water system is a testament to self-reliance, embracing independence in meeting the essential need for water. By understanding the workings of traditional and off-grid systems, we gain a broader perspective on how communities ensure access to this life-sustaining resource.

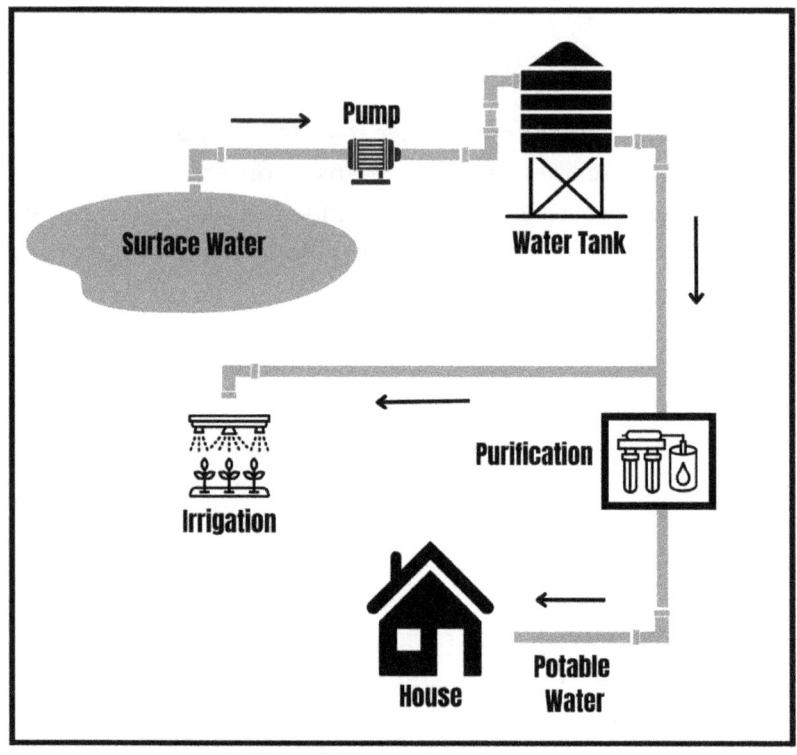

Off Grid Water Supply System

Interactive Element: Water Survival Quiz

Before you begin your journey to learn about water survival, take this quiz to assess your readiness for the challenges ahead. The quiz will evaluate your knowledge, skills, and mindset to help you navigate possible difficulties. So, get ready to find out where you stand on becoming a water survival expert. Let's begin the quiz.

1. **What is the most reliable source of drinking water in a wilderness setting?**

A. Streams and rivers

B. Puddles

C. Rainwater

D. Seawater

Answer: C. Rainwater

2. **How can you purify water in the wild without a water filter?**

A. Drinking it directly from a natural source

B. Boiling it for at least one minute

C. Letting it sit in the sun

D. Adding leaves or natural herbs

Answer: B. Boiling it for at least one minute

3. **What are the signs of dehydration?**

A. Increased energy levels

B. Dark urine

E. Decreased hunger

E. Cold skin

Answer: B. Dark Urine

4. **Which of the following should you do to minimize water loss in a survival situation?**

A. Avoid strenuous activities during the hottest part of the day

B. Drink large amounts of water at once

C. Wear heavy clothing to absorb sweat

E. Limit shelter to increase air circulation

Answer: A. Avoid strenuous activities during the hottest part of the day

5. **In a survival situation, how can you collect rainwater?**

A. Using a large leaf as a funnel

B. Digging a hole in the ground

C. Standing with your mouth open

D. Waving a piece of cloth in the air

Answer: A. Using a large leaf as a funnel

Scoring

5 Points: All correct answers

3-4 Points: Good basic knowledge

1-2 Points: Needs improvement

0 Points: Further education on water survival recommended

Wrapping it Up...

In this chapter, we have discussed the crucial significance of water survival and the role of self-sufficiency in obtaining access to clean water. We have explored the core of self-sufficiency as both a mindset and a skill set, highlighting its importance in dealing with challenges and ensuring resilience. The main takeaway from this discussion is that being prepared for water survival is not just an option; it's a necessity.

Moving ahead, Chapter 2 will build upon the foundation laid in the previous chapter by discussing different water sources and providing crucial information on collecting water from each source.

2

FIND AND COLLECT

This chapter focuses on teaching the vital skills of finding and collecting water, which are essential for survival in any environment. You will learn how to interpret nature's signals to locate hidden water sources, such as underground springs, as well as the methods for effectively collecting this valuable resource.

Groundwater

Billions of people rely on groundwater for their daily drinking water. A report recently highlighted this and revealed that 2.5 billion people depend solely on this source (Qian et al., 2020). This report tells a story that goes beyond just statistics, demonstrating the importance of groundwater for survival. It is found deep within the Earth and is crucial for sustaining life. The report shows how essential it is to recognize and protect this resource.

Groundwater Utilization

Let's explore the definition and properties of groundwater and its vital role in sustaining communities. Groundwater is the reservoir of water that exists beneath the Earth's surface, contained within permeable rocks and soils. Although it is hidden from view, its impact is profound and provides a source of life for those who depend on it daily.

Understanding groundwater dynamics depends heavily on the physical characteristics of the geological features that contain it. The permeability and porosity of the underground layers play a critical role in determining how easily water can move through them. Interconnected like the Earth's circulatory system, aquifers demonstrate this valuable resource's fluidity and adaptability.

Groundwater is defined by its physical characteristics and chemical composition, which determine its taste, quality, and suitability for different purposes. Dissolved minerals, pH levels, and the presence of contaminants provide a complete understanding of the chemical properties of groundwater. It is essential to investigate these properties to determine the groundwater's portability and potential impact on ecosystems and infrastructure (Water Science School, 2018).

Learning how to collect groundwater is an important skill that everyone should consider. Groundwater has numerous uses, including providing drinking water and supporting agriculture and industry. Due to its reliability, it is often preferred in areas where surface water is scarce or unpredictable. However, as we depend on this essential resource, it is crucial to have a comprehensive understanding of sustainable practices to ensure its longevity.

Locating Groundwater

Here are some straightforward steps to help you locate groundwater:

- *Understanding the landscape:* Begin by studying the landscape. Groundwater is more accessible in areas with permeable soil, such as sand or gravel, as it allows water to move through easily. Identifying low-lying areas or depressions can indicate potential groundwater presence.

- *Using topographic maps:* Topographic maps can be valuable tools. Look for features like valleys, where groundwater may accumulate. Understanding the topography helps in pinpointing potential well locations.

- *Consulting authorities or hydrogeological studies:* Contact local water authorities or hydrogeological surveys. They may have information on the area's geological characteristics, helping you make informed decisions on the digging process and well placement.

- *Check for previous wells*: It's important to check if there are any previous wells drilled on the property by looking at geological survey records or state well drilling reports. These records will indicate the depth of previous wells in the area and whether or not they found water.

- *Satellite Imagery:* Modern technology offers a helping hand in the form of satellite imagery from readily available mapping applications like Google Maps and aerial surveys. These help identify surface features that hint at subsurface water. Look for vegetation patterns, as thriving vegetation may indicate the presence of groundwater close to the surface.

- *Dig test wells:* Once you've narrowed down potential locations and gathered what you need, the next step is to dig test wells. These are smaller, temporary wells drilled to assess the water table and quality. Monitoring the water level over time and analyzing the characteristics of the water extracted can confirm the presence and viability of groundwater.

- *Obtain necessary permits:* Before you begin digging, check local regulations and obtain any required permits. Compliance with legal requirements ensures responsible well construction and protects the groundwater resource.

These steps will help you find groundwater for a reliable and sustainable source of fresh water.

Collecting Groundwater

Water wells have been used to access groundwater for over 8,000 years. Their construction has evolved from simple scoops of sediment in a dry watercourse to advanced systems like qanats in Persia. Modern wells are excavated by digging, driving, or drilling to access water stored in underground aquifers.

- *Digging the well:* Start drilling using the chosen method, which could involve manual labor, a hand auger, or a mechanical drilling rig (which we will discuss later). Progress through layers of soil and rock until you reach the water table. Casing pipes are inserted to prevent the well from collapsing.

- *Install screen and gravel pack:* Place a screen around the bottom of the well to prevent sediment from entering and clogging the well. A gravel pack around the screen helps stabilize the well and improve water flow.

- *Determine water quality:* As you dig, periodically test the water for quality. This step ensures that you tap into a sustainable and potable (drinkable) water source. Water testing can identify any potential contaminants.

- *Well completion:* Once the desired depth is reached, secure the well with a cap to prevent contamination and unauthorized access. The well cap also protects the water from surface pollutants. You should also consider installing a well casing, typically made of PVC or steel, to prevent the well from collapsing. It also helps protect the water from contaminants.

- *Install pump and water system:* Depending on your needs, install a pump to extract water from the well. Connect the well to a water distribution system for easy access. Ensure that the pump is suitable for the well's depth and yield.

- *Regular maintenance:* Inspect and maintain the well to prevent contamination or structural problems. Follow recommended maintenance practices to ensure the well's longevity and the extracted water's quality.

Adhere to local regulations and seek professional assistance if needed for effective groundwater collection.

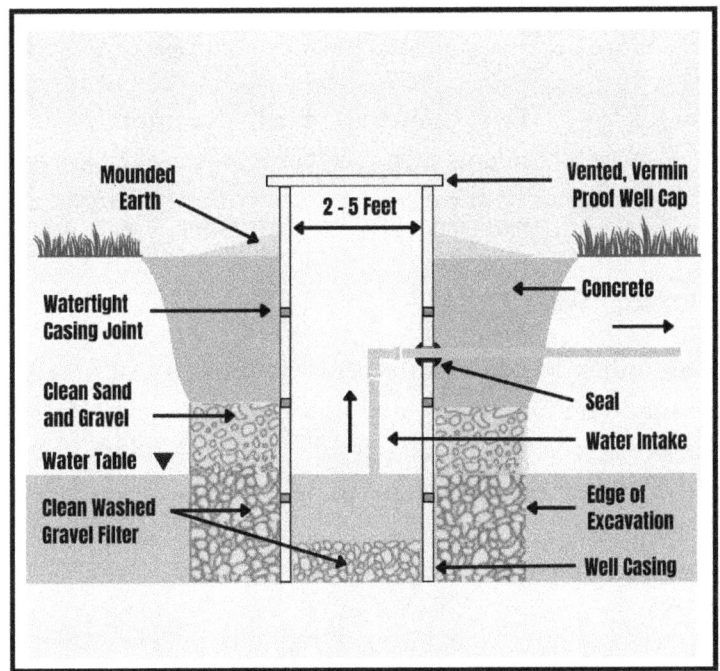

Typical Water Well Construction

Considerations:

When digging your own well, several critical considerations warrant attention:

- *Water Table Depth:* A comprehensive understanding of the water table depth in the chosen location is essential. It guides the determination of an appropriate well depth that aligns with specific needs.

- *Legal Compliance:* Navigating legal compliance is critical to responsible and lawful well construction, requiring necessary permits and adherence to local regulations.

- *Contamination:* Understanding contamination risks is also crucial, as surface runoff and proximity to septic systems can introduce pollutants. Maintaining a safe distance becomes a pivotal practice to mitigate these risks.

- *Safety at Work:* Caution is equally vital during excavation to avert accidents such as collapsing walls or falling debris, underscoring the need for a safe working environment.

Exploring Well Types

Understanding the different types of wells is crucial for collecting groundwater. Let's acquaint ourselves with the most notable methods. Each offers distinct advantages, and in off-grid survival scenarios, the simplicity of manually dug wells becomes especially valuable.

Drilled Wells

Drilled wells involve using mechanical equipment, such as drilling rigs, to bore into the Earth's subsurface. These wells can reach considerable depths and are often cased to prevent collapse. Drilled wells are effective in accessing deeper aquifers and are commonly used in both rural and urban settings. The process typically requires specialized equipment and professional expertise.

Driven Wells

Driven wells, also known as sand-point or percussion wells, are a simpler alternative. These wells are constructed by driving a casing into the ground using a driver or a hammer. The casing, typically perforated, allows water to enter the well. Driven wells are suitable for shallow aquifers and are relatively straightforward to install. They are commonly employed in areas with loose or sandy soils.

Dug Wells

Dug wells are a traditional and manual well construction method, making them particularly relevant in off-grid survival situations. These wells are excavated by hand, often using basic tools like shovels and pickaxes. Dug wells are typically shallower than drilled wells, but they can tap into vital water sources. They are cost-effective, especially when machinery may be impractical or unavailable.

Manual Well-Digging Methods

Among the methods mentioned earlier, well-digging is one that we should elaborate on. Several methods offer practical solutions for manually digging a well, each with unique advantages and disadvantages. Let's take a brief look at each process.

1. **Hand-Digging:** Using basic tools such as shovels, pickaxes, and buckets to manually excavate the soil until reaching the water table.

Advantages:

- requires minimal equipment, making it cost-effective
- suited for various terrains and accessible in remote areas

Disadvantages:

- can be physically demanding and time-consuming
- typically results in shallower wells compared to mechanized methods

2. **Augering:** This utilizes a rotating helical screw, or auger, to drill into the ground. The soil is brought to the surface as the auger turns.

Advantages:

- faster than traditional hand-digging,
- efficient for softer soils and relatively simple equipment.

Disadvantages:

- may face challenges in harder soils
- depth can be limited.

3. **Sludging**: Also known as sludge augering, this employs a sludge pump to remove debris and soil from the well during excavation.

Advantages:

- faster than traditional hand-digging

- minimizes the risk of well collapse by continuously removing excavated material

Disadvantages:

- requires a sludge pump, adding some complexity
- may have limitations in reaching deeper water tables

4. **Manual Percussion Drilling:** This involves repeatedly lifting and dropping a heavy drill bit into the well, breaking up the soil for removal.

Advantages:

- capable of reaching greater depths compared to some other manual methods
- more efficient than hand-digging alone

Disadvantages:

- still demands physical effort, albeit less than hand-digging
- requires a drilling tool and some mechanical components

5. **Drive Point:** This entails pushing a perforated pipe into the ground to form a casing for the well. Excavation then occurs within the casing.

Advantages:

- faster than traditional hand-digging
- provides immediate structural support to prevent well collapse

Disadvantages:

- may have constraints in reaching deep water tables
- requires tools for driving the casing and excavating within it

Well Digging Estimated Costs

Before digging a well, assessing the associated costs is important. Several factors, such as the depth of the well, the selected digging method, and the geological characteristics of the location, can significantly influence the overall cost. Proper evaluation of these aspects is crucial to ensuring that the well-digging project meets water needs within the allocated budget.

Water Testing

After successfully digging a well, the crucial testing phase begins. This involves analyzing the chemical composition, checking for contaminants, and confirming adherence to health standards. Testing the well water provides valuable insights into its suitability for consumption and various uses. Additionally, assessing the well's yield—how much water it can consistently produce—ensures that it meets the intended requirements. Regular testing and monitoring contribute to maintaining a reliable and safe water source, promoting the well's sustainability and longevity.

What is Rainwater Harvesting?

Rainwater harvesting is a sustainable practice that involves collecting and storing rainwater for later use. The process typically begins by directing rain from rooftops or other surfaces into a collection system, such as gutters and downspouts. This harvested rainwater is then filtered and stored in tanks or cisterns for various purposes, including irrigation, flushing toilets, or even as a potable water source.

Advantages:

- *Sustainability*: Rainwater harvesting is a sustainable water source, reducing dependence on traditional water supplies.

- *Cost savings*: It can lead to significant cost savings on water bills, especially for activities like garden irrigation or flushing toilets.

- *Environmentally friendly*: Harvesting rainwater reduces stormwater runoff, which can help prevent soil erosion and alleviate strain on local water systems.

- *Mitigates flood risks*: By capturing rainwater, the practice can contribute to reducing the risks of flooding during heavy rainfall.

Disadvantages:

- *Initial cost*: The installation of rainwater harvesting systems may involve an initial investment in equipment and setup.

- *Dependency on rainfall*: The effectiveness of rainwater harvesting is directly linked to the availability and frequency of rainfall, making it less reliable in arid regions.

- *Quality concerns*: Depending on the collection surface and storage conditions, there may be concerns about the quality of harvested rainwater. Regular maintenance and proper filtration are essential to addressing this.

Rainwater harvesting is a practical and eco-friendly method that taps into a natural resource for various uses. While it offers numerous benefits, careful consideration of factors such as location, setup costs, and reliance on rainfall should be taken into account when deciding to implement a rainwater harvesting system.

Uses of Collected Rainwater

Collected rainwater serves as a versatile resource with numerous practical applications. Some common uses include:

- *Irrigation*: Rainwater is excellent for watering plants, gardens, and agricultural crops, providing a natural and sustainable source of hydration.

- *Toilet flushing*: Harvested rainwater can be used for flushing toilets, reducing reliance on treated water for non-potable purposes. "Non-potable" means that the water isn't intended for drinking but can be used for other purposes like irrigation, industrial processes, or even replenishing aquifers (O'Donnell, 2021).

- *Outdoor cleaning*: It's suitable for tasks like washing cars, outdoor furniture, and exterior surfaces, contributing to water conservation efforts.

- *Laundry*: rainwater can be used for laundry, especially in areas where water quality standards permit its use.

- *Supplemental water supply*: In some cases, treated rainwater may even be used for drinking and cooking, following proper filtration and purification processes.

Estimating Your Rainwater Collection

The amount of rainwater you can collect depends on various factors, including the size of the collection surface, the efficiency of the harvesting system, and local rainfall patterns. A simple calculation involves determining the annual water yield using the formula:

Water Yield (gallons) = Collection Area (square feet) × rainfall (inches) × Collection Efficiency (as a decimal) × Conversion Factor (gallons per square feet)

As an example, suppose you have a roof with a collection area of 1,000 square feet, and the average annual rainfall is 30 inches. If your collection system is 80% efficient, the calculation would be:

Water Yield = 1,000sq ft × 30 in × 0.80 × 0.623g/sq ft

= 14,952 gallons

It's essential to note that this is a simplified calculation, and actual yields may vary. Factors such as evaporation, spillage, and system efficiency play roles in the actual amount of water collected.

Different Methods to Collect Rainwater

Understanding the following methods is key to making informed choices for rainwater harvesting, where each method is a step toward unlocking the potential of every raindrop.

Rain Barrels

Rain barrels represent a straightforward and accessible method of rainwater collection. These containers, often placed beneath downspouts, collect rainwater directly from rooftops. While they are simple to install and cost-effective, rain barrels are most suitable for smaller-scale residential use. However, their limited capacity and potential for water stagnation make them better suited for supplemental rather than primary water sources.

"Dry" System

The "Dry" rainwater harvesting system introduces sophistication to the collection process. It involves a network of pipes directing rainwater from rooftops to storage without any standing water in the pipes. This method is ideal for large-scale applications, such as industrial or commercial settings, where advanced filtration is crucial. While offering high-quality water, the "Dry" system demands a higher initial investment and a more intricate setup.

"Wet" System

The "Wet" rainwater harvesting system strikes a balance between simplicity and effectiveness. It includes pipes with standing water, helping to prevent freezing in colder climates. "Wet" systems are more straightforward to install and maintain than "Dry" systems, making them cost-effective and versatile. However, they may have fewer filtration mechanisms, and the standing water in the pipes introduces a potential risk of contamination. "Wet" systems are well-suited for various applications, especially in residential and smaller-scale commercial settings.

Rainwater Harvesting: A Practical Guide

A simple rain barrel systems or more complex setups can utilize rainwater as a sustainable and cost-effective resource. Follow these simple steps to create your own rainwater harvesting system:

1. Plan and Assess

    a. Assess rainfall patterns: Understand local rainfall patterns to estimate the potential water yield.

    b. Determine collection area: Identify suitable surfaces for rainwater collection, such as rooftops.

2. Calculate Water Needs

    a. Estimate usage: Determine your water needs for various purposes like irrigation, flushing, or even drinking. (An in-depth procedure for doing this will be discussed in Chapter 4.)

3. Choose a Method

    a. Select collection system: Decide on a rainwater harvesting method - be it rain barrels, a "Wet" system, or a more intricate "Dry" system.

4. Install Collection System

    a. Rain barrels: Place barrels beneath downspouts. Install a screen or filter to prevent debris from entering.

    b. "Wet" system: Set up pipes with standing water leading to storage tanks. Ensure proper filtration to maintain water quality.

    c. "Dry" system: Install pipes to direct rainwater without standing water. Include advanced filtration mechanisms.

5. Include Filtration

    a. Install filters: Regardless of the method, incorporate

filters to remove debris and impurities.

b. Consider first flush diverters: Add diverters to redirect the initial rainwater flow, reducing contaminants in the collected water. A diverter is a device designed to redirect the flow of water within a filtration system. This component serves a crucial role in guiding water through different stages of the filtration process.

Materials Needed:

- *Rain barrels*: Plastic or metal barrels, downspout diverters, screens.

- *"Wet" system*: PVC pipes, storage tanks, filters, and possibly a pump.

- *"Dry" system*: PVC pipes, storage tanks, advanced filtration system, and pumps if needed.

Cost Considerations:

- *Rain barrels*: Typically range from $50 to $200 each.

- *"Wet" system*: Costs can vary widely based on tank size, pump requirements, and filtration systems. Estimates range from a few hundred to several thousand dollars.

- *"Dry" system*: The most expensive option, with costs ranging from several thousand to tens of thousands of dollars, considering the complexity of the setup and advanced filtration.

Maintenance Tips:

- *Regular cleaning:* Keep gutters, screens, and filters clean to prevent clogs.

- *Inspect pipelines:* Check for leaks and ensure proper functioning of the entire system.

- *Monitor water quality:* Periodically test water quality, especially if it is used for drinkable purposes.

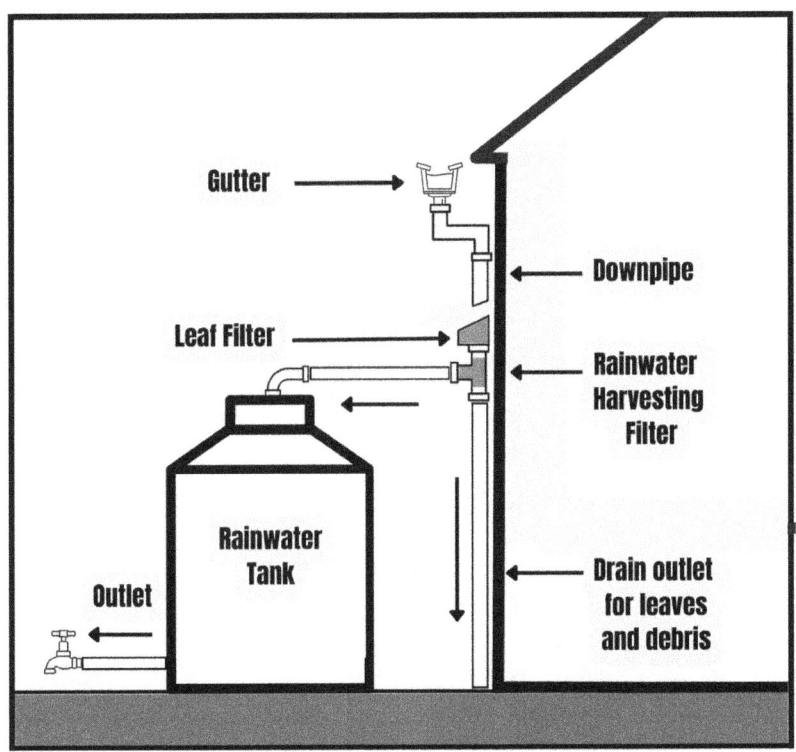

"Wet" Rainwater Harvesting System

Surface Water: Lakes, Dams, Rivers, and Streams

Surface water is the most readily available water source in the wilderness. It is essential to know how to safely identify and collect this vital resource to ensure survival.

Caution is needed however as pollution poses a significant risk to surface water. Urban and industrial areas can introduce pollutants that can negatively impact water quality. The accessibility, proximity, and environmental impact of extraction methods also determine how practical it is to use these water sources responsibly.

To make informed decisions for extraction, it is crucial to understand both the physical and chemical properties of surface water.

Lakes and Dams

Lakes, with their varying depths and surface areas, present a diverse range of water accessibility. The mineral content in lakes can influence taste and suitability for consumption, while temperature variations impact the water's dissolved gases and nutrients. Despite their potential as substantial water reservoirs, lakes are not without challenges. Contamination risks, especially in populated areas, and the potential for algae blooms to affect water quality require careful consideration.

Rivers

Rivers, characterized by flow rate, width, and depth, offer continuous water flow suitable for various applications. Factors like sediment load influencing turbidity and dissolved oxygen levels impact water quality and taste. While rivers provide reliability, the risk of contamination and varying water quality based on upstream activities necessitates vigilance in their utilization.

Streams

Smaller and often found in diverse terrains, streams possess unique characteristics influenced by gradient, width, and depth. Their accessibility makes them valuable water sources, with potential for cleaner water in less populated areas. However, streams are vulnerable to drought, and their lower water volume compared to rivers requires consideration in their utilization.

Finding Surface Water

Whether in the wilderness or exploring diverse landscapes, knowing how to locate surface water can be a lifesaver. By leveraging tools like maps, satellite imagery, and on-the-ground observations, you'll gain the insights needed to navigate and thrive in environments where surface water is essential.

- *Topographic maps*: Utilize topographic maps to identify contour lines, valleys, and depressions. Low-lying areas often indicate potential locations for surface water.

- *Satellite imagery*: Access satellite imagery to observe landscape features. Bodies of water, even small streams or ponds, are often visible from above. Tools like Apple Maps and Google Maps should help you with this.

- *Physical exploration*: Conduct on-site surveys, especially in areas with uneven terrain. Walk the landscape to visually identify signs of water flow, such as small channels or wet soil.

- *Vegetation clues*: Observe the types of vegetation in an area. Lush, green vegetation may suggest the presence of nearby water sources.

- *Animal activity*: Wildlife tends to congregate around water sources. Observe animal tracks, birds, or other signs of wildlife activity, which can lead you to surface water.

- *Ask the locals*: Engage with local communities or residents who are familiar with the area. They often possess valuable insights into the location of nearby surface water.

- *Flow drainage patterns*: Understand the natural drainage patterns of the landscape. Water tends to flow downhill, so following these patterns can lead to the discovery of streams or rivers.

- *Listen for flowing water*: Stand quietly and listen for the sound of flowing water. Even small streams can create audible sounds, guiding you to potential water sources.

- *Look for water-loving plants*: Certain plants thrive in moist environments. Identify vegetation such as cattails or willows, as they often grow near water.

- *Check soil moisture*: Damp or muddy soil indicates recent water presence. Conduct simple soil tests to assess moisture levels.

Surface Water Collection

Surface water is typically pumped from its source and screened at the intake before being stored. It can then be used for irrigation or purified for human consumption. If you have access to surface water, there are important factors to consider before deciding whether it is a viable option:

- *Regulatory Compliance:* Before starting, check with state and county agencies for water laws. Pumping in restricted areas can result in hefty fines. Contact your local water authority first. The state's water resources or environmental protection department handles water laws.

- *Water Quality:* It is essential to test the quality of surface water to ensure its safe use. This includes checking for chemical pollutants, sediment, turbidity, and metals, as well as taste and odor. Additionally, it is crucial to assess the risk of pathogens such as bacteria and viruses. If the water is intended for domestic consumption, it will require purification. Detailed information on the methods of purification can be found in Chapter 3.

- *Water Yield:* Understanding available water is important for long term use. Surface water sources can be affected by weather and withdrawals from others. This is especially true in smaller sources like ponds or streams.

Dew Water: Nightly Moisture

Dew water is the moisture that delicately forms on surfaces, such as grass and leaves, during the coolness of the night. This natural occurrence results from a simple yet fascinating process involving radiational cooling, saturation, and condensation. As the Earth's surface loses heat through radiation during the night, it cools various objects, reaching what is known as the dew point. At this temperature, the surrounding air becomes saturated with moisture, condensing water vapor into tiny droplets on cooled surfaces—reminiscent of the beads of water that form on a cold glass in a warm room (Dibley, 2022).

Several factors influence dew formation, including the night sky's clarity, humidity levels, and wind speed. Clear skies allow for more effective radiational cooling, higher humidity provides more moisture for condensation, and calm winds facilitate surface cooling.

While dew isn't a substantial water source, it has ecological significance, providing moisture to plants and participating in the local water cycle. In some conditions, such as regions facing water scarcity, people have collected dew for drinking. The formation of dew is a captivating interplay of natural elements, offering a subtle yet vital contribution to the ecosystems it touches.

Dew Harvesting Techniques

These techniques demonstrate a harmonious relationship between human ingenuity and the subtle processes of the natural world, from collecting dew on grass and leaves to harnessing in the presence of fog.

Grass Collection

One straightforward method of harvesting dew involves utilizing grass as a collection surface. During the night, dew forms on its blades as the grass cools. One can simply run a cloth or sponge along the grass to collect this dew, absorbing the moisture. This method is practical and has been employed by people in various settings for centuries.

Leaves Collection

Like grass, leaves on plants and trees provide an excellent surface for dew collection. The process involves gently brushing or wiping the leaves with a cloth or sponge to absorb the dew. This technique is effective in areas where vegetation is abundant, offering a natural and straightforward way to gather dew for various purposes.

Fog Harvesting

Fog harvesting can be an effective technique for obtaining water in areas where fog is prevalent. This method involves using specialized nets or mesh structures, often placed on elevated platforms. As the fog passes through these structures, water droplets are collected and eventually form larger droplets that can be collected and directed into containers. Fog harvesting is beneficial in areas where access to clean water is limited and in arid environments where traditional water sources are scarce.

Considerations

- *Time of day*: Dew harvesting is most effective during the night and early morning when temperatures drop, leading to condensation.

- *Vegetation density*: Areas with abundant grass and vegetation are more suitable for grass and leaf leaves collection.

- *Fog frequency*: Fog harvesting is most practical in regions with frequent foggy conditions.

While dew harvesting may provide small quantities of water, it can be a valuable supplemental source, especially in regions with limited access to traditional water supplies. These simple and nature-inspired techniques showcase the ingenuity of leveraging natural processes to meet basic water needs.

Interactive Element: Assessing Water Quality

What if you find yourself without a testing kit? This guide explores a hands-on, no-kit-needed approach to assessing water quality. From simple observations of color and clarity to engaging your senses through smell and taste, these techniques provide a resourceful means to evaluate water without relying on advanced tools.

- *Observation of color and clarity:* Start by visually inspecting the water. Clear water is generally a positive sign, while cloudiness or unusual coloration may indicate impurities or contamination.

- *Smell check: Give the water a sniff.* Foul or unusual odors could be indicative of contamination. Fresh, odorless water is generally a positive sign.

- *Taste test:* While not the most precise method, tasting the water can provide insights. Clean, freshwater should have a neutral taste. Unusual or unpleasant flavors may signal impurities.

- *Aquatic life observation:* If applicable, observe the presence and health of aquatic life in the water body. Healthy aquatic ecosystems often indicate good water quality. A lack of diverse aquatic life or the presence of distressed organisms may suggest issues.

- *Sediment inspection:* Collect a sample in a transparent container and allow it to settle. Examining the settled sediment can reveal particles and impurities. Excessive sediment may indicate poor water quality.

- *Algae presence:* Check for the presence of algae. While some algae are normal, an overgrowth, especially in vibrant colors, might indicate nutrient pollution, affecting water quality.

- *Assessment of surroundings:* Consider the environment around the water source. Urban or industrial areas nearby may increase the risk of contamination. Natural, untouched environments generally pose fewer risks.

- *Source evaluation:* Understand the source of the water. Groundwater from wells or natural springs is often purer than surface water from lakes or rivers. Knowing the source can offer clues about potential contaminants.

- *Temperature analysis:* While not a definitive indicator, temperature can provide some insights. Unusual temperature variations may be linked to industrial discharges or other contamination sources.

- *Community feedback:* Engage with local communities and residents. They may have valuable insights and historical knowledge about the water source, including any known issues.

While these methods offer preliminary insights, they are not a substitute for professional water testing. For comprehensive assessments, especially of drinking water, it is recommended that you use a certified testing kit or consult with local water authorities.

Wrapping It Up...

We have seen groundwater as an important source of water, and how the water well is a time- tested way of collecting it. Different well types and construction techniques to collect the water were offered depending on local conditions.

We have explored the important topic of surface water, including lakes, rivers, and streams. We can effectively locate and utilize these sources by understanding their physical and chemical properties. We have also learned about harvesting dew water and rainwater and practical techniques for doing so in different environments.

The upcoming chapter will explore important information regarding water purification methods. It is vital to ensure the safety and quality of the water we obtain from various sources in order to survive. Join us in discovering the techniques to purify water, which will help secure a fundamental resource for life in any situation

3

WATER PURIFICATION

In our daily lives, a subtle danger lurks in the quiet corners of our homes. The Environmental Protection Agency (EPA) warns us that once considered pure, our drinking water is now vulnerable to various potential threats, including chemicals, microbes, and radionuclides (US EPA, 2017). These threats can seep into the water we rely on for sustenance and cause harmful effects such as gastrointestinal infections, nervous system disruptions, and other internal complications.

The EPA's cautionary tale exposes a hidden narrative of danger, highlighting the importance of ensuring our drinking water is safe and free from harmful contaminants.

Let's explore the hidden crisis of compromised tap water that threatens our health and well-being. This chapter will dive into different purification methods that can help us achieve off-grid water survival. Understanding these methods is essential, empowering us to secure clean and safe water in various scenarios. This knowledge is not just a skill, but a necessity.

This guide covers everything you need to know about water purification methods. We examine each technique in detail, highlighting the pros and cons. We'll provide all the essential steps, tips, and tricks for turning information into practical knowledge. Whether you're a seasoned adventurer, a homesteader, or someone getting ready for the unexpected, these methods are the key to unlocking your water survival potential.

So, let's get right into the world of water purification. This is where every method is a stepping stone towards self-sufficiency.

Water Boiling

Boiling water is a timeless technique essential for anyone facing the uncertainties of survival. This uncomplicated method relies on the principle that heat can purify. Let's explore the importance of boiling as a crucial survival skill. Despite nature's unpredictability, this skill can significantly determine whether you are vulnerable or resilient.

The boiling method is a straightforward approach to water purification. It involves heating water to its boiling point, typically 212°F (100°C) at sea level. This process is a time-tested technique for rendering water safe for consumption by eliminating harmful pathogens, bacteria, and other contaminants (UPMC, 2017).

Advantages:

- *Accessibility:* Boiling is easily accessible to most people, requiring only a heat source and a container.

- *Pathogen elimination:* Boiling effectively kills a wide range of microorganisms, ensuring water safety.

- *Minimal equipment:* This method doesn't demand sophisticated equipment, making it feasible in various settings.

Disadvantages:

- *Time-consuming:* Boiling can be time-consuming, especially when purifying larger quantities of water.

- *Energy requirement:* It demands a consistent energy source, which might be a challenge in certain situations.

- *Doesn't remove chemicals:* While it tackles biological contaminants, boiling doesn't remove chemicals or pollutants.

Effectiveness and Duration of Boiling

The boiling method effectively eliminates bacteria, viruses, and parasites. It is a reliable first line of defense against waterborne diseases, particularly in areas with uncertain water quality.

Boiling water does not remove chemical pollutants, heavy metals, or dissolved solids (Kristanti et al., 2022).

Boil water for one minute for safety. At higher altitudes, boil for three minutes. Boiling water is practical in emergencies. It kills harmful microorganisms but has its limits against chemical contaminants. Remember to be patient and let the water reach and maintain a rolling boil for the prescribed duration to guarantee it's safe for consumption.

Distillation

Distillation is a reliable method for water purification that operates on the principle of separating components based on their different boiling points. It involves heating water to its boiling point, collecting the vapor, and then condensing it into liquid form. This process effectively removes impurities, ensuring a purified water supply.

Advantages:

- *Comprehensive purification:* Distillation is highly effective in removing many contaminants, including bacteria, viruses, and dissolved minerals.

- *Chemical-free:* Unlike some other methods, distillation doesn't rely on chemicals, making the purified water free from added substances.

- *Versatility:* It can be applied to various water sources, providing a versatile solution for different environments.

Disadvantages:

- *Energy intensive:* Distillation requires a significant amount of energy, which may be a limiting factor in certain situations.

- *Slow process:* The distillation process can be time-consuming, particularly when purifying large quantities of water.

- *Limited removal of volatile compounds:* It may not effectively remove certain volatile compounds with boiling points close to that of water.

Effectiveness and Duration of Distillation

Distillation is a highly effective method for removing a broad spectrum of contaminants. It excels at eliminating bacteria, viruses, heavy metals, and other impurities, providing a thorough purification process.

Distillation is particularly efficient in separating substances with boiling points distinct from that of water.

The time required for distillation varies based on factors such as the quantity of water and the efficiency of the distillation apparatus.

Generally, it is a slower process compared to some other purification methods.

Water Distillation Steps:

- *Boiling:* Heat the water to its boiling point.
- *Vapor Collection:* Capture the vapor released during boiling.
- *Condensation:* Convert the vapor back into liquid form by cooling it.
- *Collection:* Gather the condensed liquid, which is now purified water.

Distillation offers a robust solution for water purification, ensuring the comprehensive removal of contaminants. While it demands energy and time, its effectiveness makes it a valuable tool when water quality is uncertain. Understanding the steps involved in distillation equips individuals with a practical skill for securing a clean and safe water source.

Chlorination

Chlorination is a widely adopted method for purifying water, leveraging chlorine's disinfectant properties. In this process, chlorine or chlorine compounds are added to water to eliminate bacteria, viruses, and other harmful microorganisms, rendering the water safe for consumption.

Advantages:

- *Broad-spectrum disinfection:* Chlorination effectively targets various microorganisms, ensuring comprehensive water disinfection.

- *Cost-effective:* It is a relatively cost-effective method, making it accessible for large-scale water treatment.

- *Long-lasting residual effect:* Chlorine residuals can persist in water, providing continued protection against microbial recontamination.

Disadvantages:

- *Chemical by-products:* Chlorine reacting with organic matter in water can produce disinfection by-products, some of which may be harmful in high concentrations.

- *Limited effectiveness against some pathogens:* Chlorination may not effectively eradicate certain parasites and protozoa, single-celled microscopic organisms in the eukaryotic domain. Some of these organisms are known to cause diseases.

- *Taste and odor:* The addition of chlorine can alter the taste and odor of water, which may be a concern for some individuals.

Effectiveness and Duration of Chlorination

Chlorination is highly effective in disinfecting water and eliminating pathogenic microorganisms. It is a trusted method for ensuring water safety in various settings.

Chlorination primarily targets biological contaminants but is less effective in removing chemical pollutants, heavy metals, or other non-biological impurities..

The time required for chlorination varies depending on the water temperature and the specific chlorine compound used. Generally, a contact time of 30 minutes to several hours is recommended for effective disinfection..

Chlorination DIY

- *Select chlorine source:* Choose a suitable chlorine source, such as household bleach containing 5-6% sodium hypochlorite.

- *Measure proper dosage:* Follow guidelines for the correct amount of chlorine to add based on the water volume. Typically, a few drops per gallon are sufficient.

- *Mix thoroughly:* Ensure even distribution by stirring or shaking the water after adding chlorine.

- *Wait for contact time:* Allow the chlorine to act on the water for the recommended contact time.

- *Check for residual smell:* A faint chlorine smell indicates proper disinfection.

Chlorination provides a pragmatic and accessible means of water disinfection. While effective, users should be mindful of potential by-products and taste alterations. By understanding the simple steps of chlorination, individuals can confidently apply this method to enhance the safety of their water supply in various situations.

Filtration

Filtration is a straightforward water purification method that relies on physical barriers to remove impurities. It involves passing water through a medium—like sand, charcoal, or specialized filters

- to trap particles, bacteria, and other contaminants, resulting in cleaner and safer water.

Advantages:

- *Versatility:* Filtration is adaptable to various settings and water sources, making it a versatile purification method.

- *Ease of use:* It's user-friendly and doesn't require specialized knowledge or equipment.

- *Cost-effective:* Many filtration methods are affordable and accessible.

Disadvantages:

- *Limited effectiveness against some contaminants:* Filtration may not effectively remove dissolved salts, heavy metals, or certain chemical pollutants.

- *Maintenance:* Filters need periodic replacement or cleaning to maintain efficiency.

- *Potential for bacterial growth: In some cases, stagnant water in filters can become a breeding ground for bacteria.*

Effectiveness and Duration of Filtration

Filtration removes physical impurities such as sand, sediment, and larger particles. Depending on the pore size of the filter, it can also capture bacteria and parasites.

Filtration is less effective in removing dissolved chemicals or substances with smaller particle sizes.

The time it takes for filtration depends on the method and the volume of water being treated. Gravity filters may take longer, while some advanced filters can provide rapid purification.

Making Your Own Water Filtration System

- *Gather materials:* You'll need a container, a filter medium (sand, gravel, activated charcoal), and a cloth or fine mesh.

- *Create layers:* Place layers of the filter medium in the container, with the finest material on the bottom.

- *Secure the filter:* Use a cloth or mesh at the bottom to secure the medium.

- *Pour water through:* Pour water into the container, and let it percolate through the layers.

- *Collect filtered water:* The water that passes through the layers is now filtered and ready for use.

Filtration offers a practical and accessible approach to water purification. While effective for many impurities, it's essential to understand its limitations. Crafting a simple filtration system at home can be a useful skill, especially in situations where clean water sources are scarce.

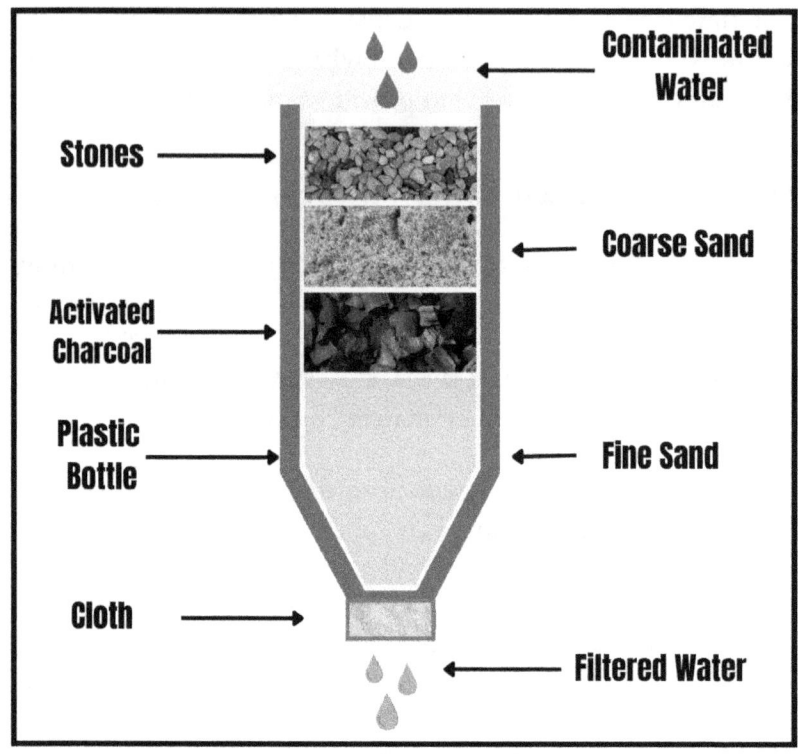

Homemade Water Filter

Activated Carbon Filtration

Carbon filtration is a practical and effective method for improving water quality, especially regarding taste and odor. Although it has several advantages, it is critical to understand its limitations and periodic maintenance to ensure lasting effectiveness. Crafting a basic carbon filter at home can be a hands-on solution for those who want a do-it-yourself approach to obtaining cleaner water.

Advantages:

- *Effective odor and taste removal:* Carbon filtration eliminates unpleasant tastes and odors from water.

- *Versatile contaminant removal:* It can effectively remove chlorine, certain chemicals, and some microorganisms.

- *Long-lasting filters:* Activated carbon filters often have a reasonable lifespan before requiring replacement.

Disadvantages:

- *Limited against some contaminants:* While excellent for taste and odor, carbon filtration may not be as effective *against minerals, salts, or some dissolved substances.*

- *Regular replacement:* Filters need periodic replacement to maintain efficiency.

- Initial cost: Some carbon filtration systems can have a higher initial cost.

Effectiveness of Carbon Filtration

Activated carbon filtration is particularly effective in improving the taste and odor of water. It can also remove chlorine, some organic chemicals, and certain microorganisms, providing a valuable defense against common contaminants. Carbon filtration may not effectively eliminate minerals, salts, or heavy metals.

Making Your Own Carbon Filter

- *Gather materials:* You'll need activated carbon, a container, a fine mesh or cloth, and a means to secure the carbon.

- *Create layers:* Place a layer of activated carbon in the container, covering it with a fine mesh or cloth to prevent particles from escaping.

- *Secure the filter:* Ensure the carbon is securely held in place, and the water can pass through it.

- *Test your filter:* Pour water through the filter and observe the improvements in taste and odor.

- *Replace as needed:* Over time, the effectiveness of the carbon diminishes, so periodic replacement is essential for optimal results.

Carbon filtration is a practical and effective method for enhancing water quality, particularly in terms of taste and odor. While it has strengths, understanding its limitations and periodic maintenance is crucial for sustained efficacy. Crafting a simple carbon filter at home can be a hands-on solution for those seeking a DIY approach to cleaner water.

Reverse Osmosis

Reverse osmosis (RO) is a meticulous water purification process that relies on a semi-permeable membrane to remove impurities. This method forces water through the membrane, allowing only water molecules to pass through. At the same time, contaminants are left behind and flushed away. The result is purified water of exceptional quality.

Advantages:

- *Comprehensive filtration:* RO effectively removes many impurities, including dissolved salts, minerals, and contaminants.

- *High purity:* It produces water with a high degree of purity, making it suitable for various applications, including drinking water.

- *Compact and versatile:* RO systems are available in various sizes and fit diverse settings, from households to industrial applications.

Disadvantages:

- *Wastewater production:* RO systems generate wastewater as a by-product, and the process can be water-intensive.

- *Energy consumption:* The method requires energy to push water through the membrane, contributing to its operational cost.

- *Limited against some contaminants:* While effective against many impurities, RO may not remove certain volatile organic compounds and gases.

Effectiveness and Duration of Reverse Osmosis

Reverse osmosis effectively removes a broad spectrum of contaminants, including bacteria, viruses, dissolved minerals, and salts. It surpasses standard filtration methods by providing a finer level of purification.

The time it takes for reverse osmosis depends on the system's capacity and water pressure. Generally, it can produce purified water at a rate of a few gallons per hour.

DIY Reverse Osmosis Filtration System

Creating a DIY reverse osmosis system is challenging due to the complexity of the technology. Commercially available RO systems are designed for optimal performance and safety. However, if you are determined to proceed:

- *Research components:* Study the components needed, including a semi-permeable membrane, pressure pump, and pre-filters.

- *Assemble carefully:* Assemble the components carefully, ensuring airtight connections.

- *Follow guidelines:* Adhere to manufacturer guidelines for membrane installation and system operation.

- *Test and monitor:* Regularly test the water produced and monitor system performance. Adjust parameters as needed.

Reverse osmosis is a powerful method for achieving water purity. While it has drawbacks, particularly regarding energy consumption and wastewater production, its ability to deliver exceptionally clean water makes it a favored choice in various applications. For those contemplating a DIY approach, thorough research and adherence to guidelines are crucial for success.

Commercial Four Stage Reverse Osmosis Kit

UV Treatment

UV treatment is a water purification method that utilizes ultraviolet (UV) light to neutralize or destroy microorganisms. In this process, water passes through a chamber where UV lamps emit light, disrupting the DNA of bacteria, viruses, and other pathogens, rendering them incapable of reproducing and causing harm (Woodard, 2019).

Advantages:

- *Microbial elimination:* UV treatment is highly effective at disinfecting water, eliminating many bacteria, viruses, and parasites.

- *No chemical additives:* It doesn't involve using chemicals, ensuring the treated water remains free from added substances.

- *Minimal maintenance:* UV systems generally require minimal maintenance, with periodic lamp replacement being the primary concern.

Disadvantages:

- *Ineffectiveness against non-biological contaminants:* UV treatment does not remove physical impurities, chemicals, or heavy metals. It solely addresses microbial concerns.

- *Dependency on clarity:* Water must be clear for UV light to penetrate effectively. Turbid or cloudy water can reduce its efficiency.

- *Electrical dependency:* UV systems rely on electricity, which may be a limitation in certain situations.

Effectiveness and Duration of UV Treatment

UV treatment neutralizes bacteria, viruses, and parasites, making it a reliable method for disinfecting water. However, its scope is limited to microbial contaminants and doesn't address chemical or physical impurities.

Water flowing through a UV chamber only takes 10-20 seconds of exposure to the lamps to kill bacteria and waterbourne microbes.

Estimated Cost of a UV Water Filtration System

The cost of a UV water filtration system can vary based on factors like brand, capacity, and additional features. Generally, residential UV systems can range from a few hundred to a few thousand dollars. It's essential to consider the initial cost and ongoing expenses, such as replacement lamps and maintenance.

UV treatment is a potent solution for microbial disinfection, particularly when chemical-free purification is preferred. While it doesn't address all contaminants, its efficiency against microorganisms makes it a valuable addition to water treatment strategies. Weighing the initial cost against long-term benefits is crucial for making an informed decision about a UV system.

Solar Disinfection

Also known as SODIS, solar disinfection is a straightforward and sustainable water purification technique that utilizes sunlight to eliminate harmful microorganisms. In this method, water is exposed to sunlight in clear plastic or glass containers, and the combined effect of ultraviolet (UV) radiation and heat works to deactivate and kill pathogens in the water (Marugán et al., 2020).

Advantages:

- *Cost-free energy source:* Solar disinfection relies on sunlight, making it a cost-free and environmentally friendly method.

- *Simple and accessible:* The technique is easy to understand and implement, requiring minimal equipment.

- *Microbial reduction:* SODIS effectively reduces the concentration of bacteria, viruses, and parasites in water.

Disadvantages:

- *Limited to microbial disinfection:* Solar disinfection addresses microbial contaminants and does not remove chemical or physical impurities.

- *Dependence on weather conditions:* Its effectiveness can be influenced by weather conditions, as overcast skies or inadequate sunlight will hinder the process.

- *Time-consuming:* To ensure thorough disinfection, SODIS requires extended exposure times, often up to six hours or more.

Effectiveness and Duration of Solar Disinfection

Solar disinfection effectively reduces the microbial load in water, particularly when exposure to sunlight is prolonged. It is most reliable for disinfecting clear water with low turbidity.

Solar disinfection does not remove chemicals, minerals, or physical impurities.

Making Your Own Solar Purifier

- *Choose a container:* Select a clear glass or plastic container with a tight-fitting lid.

- *Fill with water:* Fill the container with the water to be treated, leaving some airspace.

- *Place in sunlight:* Put the container in direct sunlight, preferably on a reflective surface to enhance UV exposure.

- *Expose for adequate time:* Allow the water to be exposed to sunlight for at least six hours, or longer on cloudy days.

Solar disinfection provides a simple and accessible means of purifying water using the sun's natural energy. While it may not address all types of contaminants and requires careful consideration of weather conditions, its cost-free nature and effectiveness against microorganisms make it a valuable tool, especially in resource-limited environments. Creating a solar purifier at home involves minimal steps, emphasizing its practicality and ease of use.

Maintaining Purification Systems

Ensuring the consistent and reliable operation of water purification systems is paramount for safeguarding the quality of your drinking water. Regular maintenance prolongs the system's life and preserves its efficiency in removing contaminants. Neglecting maintenance can lead to compromised water quality and the potential for health risks.

Maintenance Techniques

Regular Inspections:

- *Why:* Routine visual checks can identify issues early on, preventing major malfunctions.

- *How:* Look for leaks, cracks, or any visible damage in pipes, filters, and other system components.

Filter Replacement:

- *Why:* Filters play a crucial role in trapping impurities. Over time, they can become clogged, reducing effectiveness.

- *How:* Follow the manufacturer's guidelines for filter replacement schedules. Regularly replace or clean filters as recommended.

Sanitization:

- *Why:* Bacteria and algae can build up in water systems, affecting water quality.

- *How:* Use appropriate sanitizing agents or follow recommended cleaning procedures to eliminate microbial growth.

Pressure Checks:

- *Why:* Proper pressure is essential for some purification methods like reverse osmosis.

- *How: Monitor pressure gauges regularly and address any deviations from the recommended levels promptly.*

UV Lamp Replacement:

- *Why:* In UV treatment systems, the UV lamp's effectiveness diminishes over time.

- *How:* Follow the manufacturer's instructions for UV lamp replacement intervals. Ensure timely replacement to maintain disinfection efficiency.

Seal and Gasket Inspection:

- *Why:* Leaks compromise the integrity of the system and can introduce contaminants.

- *How:* Regularly check and replace damaged seals and gaskets. Ensure proper sealing in all components.

Monitoring Water Quality:

- *Why:* Periodically test the treated water to ensure the purification system is operating as intended.

- *How:* Use water testing kits to assess key parameters like microbial content, pH, and chemical concentrations.

Professional Servicing:

- *Why:* Periodic professional maintenance can address complex issues and ensure optimal system performance.

- *How:* Schedule regular servicing by qualified technicians, especially for more intricate systems like reverse osmosis.

Documentation:

- *Why:* Keeping records of maintenance activities helps track the system's performance and informs future maintenance needs.

- *How:* Maintain a log that includes dates of filter changes, inspections, and any repairs undertaken.

Consistent maintenance is key to the reliability of water purification systems. A proactive approach, including regular inspections, timely component replacements, and adherence to recommended guidelines, ensures that your purification system remains a steadfast guardian of water quality.

Interactive Element: Purification Options Table

In this comprehensive overview, we present various techniques for purifying water, each with its own set of pros and cons. Whether you're an outdoor enthusiast, a home DIYer, or simply curious about ensuring the safety of your drinking water, this interactive element breaks down the essentials.

Navigate through the table to discover the advantages and disadvantages of each method, understand what chemicals or components they target, and find step-by-step action plans for implementation. From boiling to solar disinfection, empower yourself with the knowledge to make informed decisions about the best purification method that suits your needs.

Method	Pros and Cons	Chemicals/ Components Removed	Action Steps	Indicative Cost
Reverse Osmosis	*Pros:* Comprehensive filtration. *Cons:* Wastewater production, Energy-intensive	*Removes:* Bacteria, viruses, dissolved salts.	Use the RO system, consider maintenance.	$300-$3,000
Distillation	*Pros:* Comprehensive purification. *Cons:* Energy-intensive.	*Removes:* Bacteria, viruses, heavy metals.	Boil water, collect vapor, condense.	$200-$2,000
UV Treatment	*Pros:* Microbial elimination. *Cons:* Ineffective against non-biological contaminants.	*Removes:* Bacteria, viruses.	Expose water to UV light, monitor clarity.	$50-$500
Carbon Filtration	*Pros:* Effective taste and odor removal. *Cons:* Limited against certain contaminants.	*Removes:* Chlorine, some organic chemicals.	Use activated carbon, replace filters regularly.	$50-$500
Chlorination	*Pros:* Broad-spectrum disinfection. *Cons:* Chemical by-products.	*Removes:* Bacteria, viruses.	Add chlorine, mix, wait, check for residual smell.	$20-$200
Filtration	*Pros:* Versatile, user-friendly. *Cons:* Limited against some contaminants.	*Removes:* Physical impurities, some bacteria.	Choose a filter medium, assemble layers, pour water.	$30-$300
Boiling	*Pros:* Simple, effective against microbes. *Cons:* Time-consuming.	*Removes:* Bacteria, viruses.	Boil water for at least 1-3 minutes.	Minimal - Gas/Electricity
Solar Disinfection	*Pros:* Cost-free, microbial reduction. *Cons:* Weather-dependent, time-consuming.	*Removes:* Bacteria, viruses.	Choose a clear container, expose it to sunlight for 6+ hours.	Container and sunlight

Note: Estimated costs are indicative and may vary based on specific brands, models, and local market conditions. Costs and amounts are in USD.

Wrapping it up...

If you are living off-grid, it is crucial to have a dependable water purification system that guarantees access to clean water for your health and well-being. This chapter explored various purification methods, each with advantages and considerations. From the straightforward process of boiling to the thoroughness of reverse osmosis, you have multiple choices that can be tailored to your specific requirements and resources.

Each method has merits when considering cost-effectiveness, convenience, and efficiency. Boiling stands out for its simplicity, while distillation excels in comprehensive purification. Chlorination offers a balance, and filtration methods like carbon and reverse osmosis provide versatility. UV treatment and solar disinfection cater to those who value eco-friendly solutions.

When deciding on the best method, consider cost, accessibility, and personal preferences. There is no one-size-fits-all solution, as your choice depends on your unique needs and available resources. You should consider adopting more than one purification method as it provides flexibility and redundancy to your water survival plan.

As you consider your path to purification, look ahead to the next chapter, where we will discuss the importance of reliable water storage. Having a dependable way to store sources of drinkable water is crucial to achieving self-sufficiency, and we will provide practical insights to complement your water purification efforts.

4

LONG TERM STORAGE

Water is essential for life but is often carelessly managed, which can have severe consequences. The World Bank has emphasized the importance of purposeful design in creating water storage solutions to ensure resilience, sustainability, and, most importantly, to save lives (Loucks & van Beek, 2017).

From ancient civilizations that depended on rivers to modern cities with complex water infrastructure, water management has always been integral to human progress. This chapter explores the challenges, choices, and outcomes of water storage. It explains why purposeful design is crucial for a sustainable and resilient future.

Your Perfect Water Storage Container

In off-grid living scenarios, where the conventional luxuries of constant water supply might be a distant dream, choosing a water storage container becomes a pivotal decision. A few key factors can make a big difference when selecting the ideal storage container.

Let's break it down into simple terms:

Capacity

Think about how much water you need. If you're in a big household or planning for emergencies, consider a container with a larger capacity. On the other hand, for solo adventurers or small spaces, a smaller container is suitable. It's like choosing the right backpack size for your journey—you don't want it too big or too small.

Durability

Your water container must be tough, especially if you're in it for the long haul. Consider where you'll be storing it—indoors, outdoors, or maybe taking it on outdoor adventures. Look for materials that can handle a bit of rough and tumble without springing a leak.

Shape

The shape of your water container can have a more significant impact than you think. A tall, slim container can be an excellent space-saving solution with limited storage space. However, a wider container might be more convenient for those with more space available. It's essential to find a container that fits your storage space properly.

Material

Choose a suitable material for your water container based on its intended use and safety for storing drinking water. Different materials have different properties, so consider the characteristics you need for your specific purpose.

Color

Did you know that the color of your water container may not just be for visual purposes? In reality, it can have an impact on the water inside. If you want to keep your water cool, choose a container in a lighter color. This is because lighter colors reflect heat, which can help maintain a cooler temperature. Conversely, darker colors can absorb more heat, which is excellent if you want warm water.

So, before you choose a water storage container, consider all these factors and find one that suits your needs, withstands challenges, and seamlessly fits into your space.

Calculating Water Usage

Knowing how much water your household consumes empowers you to make informed decisions for conservation. It prepares you for potential challenges like water shortages or disruptions. Calculating your water usage and needs doesn't have to be complicated. Here's a ten-step guide to help you figure it out:

1. *Identify water sources*: Note all the water sources in your home, including taps, toilets, washing machines, and outdoor water use like gardening.

2. *Track water usage*: Over a set period, say a week, monitor and record your water usage. Check your water bills for usage details, or use a water meter if available. Keep an eye on activities like showers, laundry, and dishwashing.

3. *Convert measurements*: Convert your water usage to a standard measurement, such as gallons or liters. This step ensures consistency in your calculations.

4. *Estimate daily usage*: Divide the total water usage by the

number of days you tracked. This gives you an average daily water usage. For example, if you used 700 gallons in a week, your daily average would be 100 gallons (700 / 7 days).

5. *Assess efficiency*: Evaluate your water fixtures and appliances for efficiency. Are there leaky faucets or toilets? Upgrading to water-efficient appliances can make a big difference.

6. *Factor in outdoor use*: If you have a garden or lawn, consider the water used for irrigation. This is often a significant part of household water consumption.

7. *Consider local factors*: Consider your local climate and water availability. Water needs may be higher in arid regions, and conservation becomes crucial.

8. *Plan for emergencies*: Factor in emergency scenarios. What if there's a disruption in your water supply? Having an emergency water reserve can be a smart move.

9. *Adjust for family size*: Consider the number of people in your household. Generally, each person may need around 70–100 gallons of daily water for all activities.

10. *Plan for the future*: Anticipate changes in your household, such as additions or lifestyle changes. This ensures your water system remains adequate for your evolving needs.

With these steps, you'll gain a good understanding of your household's water usage and needs. By gaining this insight, you become a steward of water, ensuring that every drop is used wisely and that your household remains resilient in the face of changing circumstances.

Long-Term Storage Containers

The significance of selecting the proper containers cannot be overstated. It's not merely about holding water; it's about ensuring that your efforts translate into a reliable and secure water supply. A misstep in container choice can compromise the integrity of your stored water, making all your meticulous planning seem like an exercise in futility.

It is crucial to select suitable containers for long-term water storage. Investing in high-quality containers ensures that your efforts to store water are purposeful and long-lasting.

Plastic Drums or Barrels

Let's start with the basics. Plastic drums or barrels are sturdy cylindrical containers made from plastic, typically high-density polyethylene (HDPE). They come in various sizes and are designed to store and transport liquids, making them versatile for different applications..

Benefits:

- Plastic drums are tough cookies. They can withstand rough handling, harsh weather, and are less prone to rust compared to metal containers. This durability ensures that your water stays secure.

- Unlike their heavier counterparts, plastic drums are relatively lightweight, making them easier to move around. This is especially handy if you need a portable water storage solution for camping or emergencies.

- Plastic drums come in a range of sizes, catering to different needs. Whether you're a solo adventurer or preparing for a family, there's likely a size that fits your water storage requirements.

- Plastic drums are often more budget-friendly than other types of containers. This makes them an economical choice for those looking for efficient water storage without breaking the bank.

Does Water Ever Go Bad in Plastic Drums?

The short answer is no; water doesn't go bad in plastic drums per se. However, the container's cleanliness and the quality of the water you initially store matter. Over time, the stored water might taste off or stale, but it is still safe to drink. If the water is properly treated and the drums are kept clean and sealed, water can be stored for extended periods without significant deterioration.

Selecting a Water Barrel According to Size

- *Assess your needs*: Consider the number of people in your household and your typical water usage. This will help determine the size of the barrel you need.

- *Available space:* Think about where you plan to store the barrel. If space is limited, a smaller, more compact size might be better.

- *Future planning*: Anticipate any changes in your water needs, such as an increase in family size or extended periods of use. Choosing a slightly larger barrel allows for future flexibility.

Estimated Cost of Plastic Barrels:

Prices for plastic drums or barrels vary based on size, brand, and additional features. On average, smaller barrels cost around $30–$50, while larger ones may range from $60–$150 or more.

Balancing your budget with the capacity and durability you require is essential.

Stainless Steel Containers

Imagine a water container with elegance and durability–that's the realm of stainless steel containers. These are sleek, metallic vessels crafted from stainless steel, a corrosion-resistant alloy known for strength and hygiene.

Benefits:

- Stainless steel containers are built to last. They resist corrosion, rust, and dents, making them sturdy companions for the long haul. This durability ensures that your water stays pristine, even in challenging conditions.

- One of stainless steel's standout features is its nonreactive nature. It doesn't leach harmful substances into water, ensuring you drink pure water without unwanted additives. Plus, its smooth surface is easy to clean, promoting excellent hygiene.

- Stainless steel has impressive thermal properties. Compared to other materials, it can keep your water cooler or warmer for longer, adding versatility to your hydration experience.

- If you're environmentally conscious, stainless steel is a green choice. It's fully recyclable, reducing the environmental impact. Choosing stainless steel aligns with sustainability efforts, making it a responsible option.

Plastic vs. Stainless Steel Containers

Durability:

- *Plastic*: Prone to scratches, dents, and can degrade over time.

- *Stainless steel*: Resistant to dents and corrosion and maintains its sleek appearance for longer.

Hygiene:

- *Plastic*: May leach chemicals into water over long periods, especially in prolonged exposure to sunlight.

- *Stainless steel:* Non-reactive, ensuring water purity without any taste or odor interference.

Environmental Impact:

- *Plastic*: Non-biodegradable and can contribute to environmental pollution.

- *Stainless steel:* Fully recyclable, reducing long-term environmental impact.

Temperature Regulation:

- *Plastic*: Less effective in maintaining water temperature.

- *Stainless steel:* Excellent thermal properties, keeping the water cooler or warmer as desired.

Estimated Cost of Stainless Steel Containers:

Compared to their plastic counterparts, stainless steel containers often have a slightly higher price tag. Depending on size, brand,

and additional features, they cost $20–$100 or more. However, the longevity and quality of stainless steel offset the initial investment.

Glass Bottles and Jars

Picture your favorite beverage or preserved treat held in a clear, sturdy container—that's the magic of glass bottles or jars. These containers, as the name suggests, are crafted from glass, a timeless material known for its clarity, purity, and recyclability.

Benefits:

- Glass is non-porous and doesn't leach harmful chemicals into your liquids. When you store water in a glass container, you get the pure, untainted taste of your beverage without any interference from the container.

- Unlike some other materials, glass doesn't absorb odors or flavors. This means your water will taste like water, not whatever was stored in the container.

- Glass is a champion in the eco-friendly arena. It's fully recyclable, and recycling glass requires less energy than producing new glass, making it a sustainable choice.

- Glass bottles or jars can withstand scratches and won't deteriorate over time. They're also less prone to discoloration, maintaining their crystal-clear appearance for an extended period.

Estimated Cost of Glass Bottles and Jars:

The cost of glass bottles or jars can vary based on size, design, and brand. On average, you might pay $5–$20 or more per bottle or jar. While glass containers might have a slightly higher initial cost, their durability and reusability can make them cost-effective.

Plastic vs. Glass Bottles

Purity:

- *Plastic*: Can leach chemicals into the contents over time.

- *Glass*: Non-porous and doesn't impart any taste or odor to the stored liquid.

Recyclability:

- *Plastic*: Some plastics can be recycled, but the process is more complex and energy-intensive.

- *Glass*: Fully recyclable, and recycling glass is a more straightforward and environmentally friendly process.

Durability:

- *Plastic*: Prone to scratches, cracks, and degradation over time.

- *Glass*: Resistant to scratches, maintains its clarity, and lasts longer.

Environmental Impact:

- *Plastic*: Non-biodegradable and can contribute to pollution.

- *Glass*: Environmentally friendly, with a lower environmental impact in the long run.

Water Bladders

Imagine a water container as flexible as your plans—that's the essence of water bladders. These containers are like portable, collapsible bags designed to hold water. Sometimes called hydration packs, they're made from durable, lightweight materials. They can be easily stored or carried, offering a versatile solution for on-the-go hydration. Large-capacity bladders are also available and a viable alternative to water tanks.

Benefits:

- Water bladders are the nomads of the water container world. They are incredibly lightweight and can be folded or rolled up when empty, making them an ideal choice for camping, hiking, or any adventure where space is at a premium.

- When not in use, water bladders take up minimal space. This is a game-changer for those who need a compact water storage solution, whether it's for a backpacking trip or emergency preparedness.

- These bladders are versatile. You can use them for drinking water on a hike, as an additional reservoir for your camping setup, or even as an emergency water supply. Their adaptability makes them a handy tool in various situations.

- Despite their flexibility, water bladders are built to last. They are often made from robust materials that resist punctures and leaks, ensuring your water stays secure, even in rugged outdoor environments.

- Large-capacity water bladders can hold up to 210,000 gallons and are used for industrial or water utility purposes. Smaller-capacity bladders of 100 gallons or more can be used for household purposes. These bladders are easy to transport when empty and can fit into confined spaces, such as underneath a house or porch.

Estimated Cost of Water Bladders:

Water bladders come in different sizes and brands. Smaller bladders typically cost between $10 and $200. These small bladders are practical and affordable and are preferred by those who need a flexible, portable water storage solution. For larger capacity bladders of 100 gallons or more, prices start at around $500, with the cost per gallon decreasing as the capacity increases.

Clay Pots

Crafted by skilled artisans, clay pots have been a cultural staple for centuries. Their porous nature allows a unique interaction with water, creating a natural cooling effect through evaporation.

Benefits:

- Clay pots have a magical way of keeping water cool. Water seeps out of the pot's tiny pores and evaporates, creating a cooling effect. It's like having a natural refrigerator for your water.

- The porous nature of clay allows it to impart a subtle earthy flavor to the water. Many people appreciate this distinctive taste, finding it refreshing and a departure from the neutral taste of other materials.

- As water interacts with the clay, it can absorb minerals from the pot, adding a touch of natural goodness to your hydration. This mineral infusion is often considered a health benefit.

- Clay is a natural material from the earth, making clay pots an eco-friendly choice. When their lifespan ends, they can return to the planet, completing a sustainable cycle.

Estimated Cost of Clay Pots:

The cost of clay pots can vary based on size, design intricacy, and craftsmanship. On average, you might find them in the range of $10–$50 or more per pot. While they may have a higher upfront cost than some other materials, their unique benefits make them a distinctive and worthwhile investment.

Water Tanks

Water tanks are like the superheroes of water storage—giant, robust containers that hold substantial amounts of water. They come in various shapes and sizes, from smaller ones for residential use to massive industrial or community needs tanks. Water tanks are typically made from plastic, fiberglass, or metal.

Benefits:

- The primary advantage of water tanks is their capacity. They can hold a significant volume of water, ensuring you have a stable and ample supply for various needs, whether for your home, farm, or community.

- Many water tanks are designed to collect rainwater. This eco-friendly feature allows you to harness nature's gift, reduce reliance on external water sources, and contribute to sustainability.

- Having a water tank is like having a reserve for unexpected situations. In emergencies or during water shortages, having a stored water supply can be a lifeline, ensuring you're prepared for the unexpected.

- Water tanks are versatile. Whether you need water for domestic use, irrigation, or industrial processes, there's a tank size and type that can cater to your specific needs.

- Once installed, water tanks require minimal maintenance. Regular cleaning and checks ensure the stored water remains in top-notch condition.

Depending on your location and water usage, having a water tank can lead to cost savings over time. It allows you to manage your water consumption efficiently and, in some cases, may even qualify for incentives or rebates.

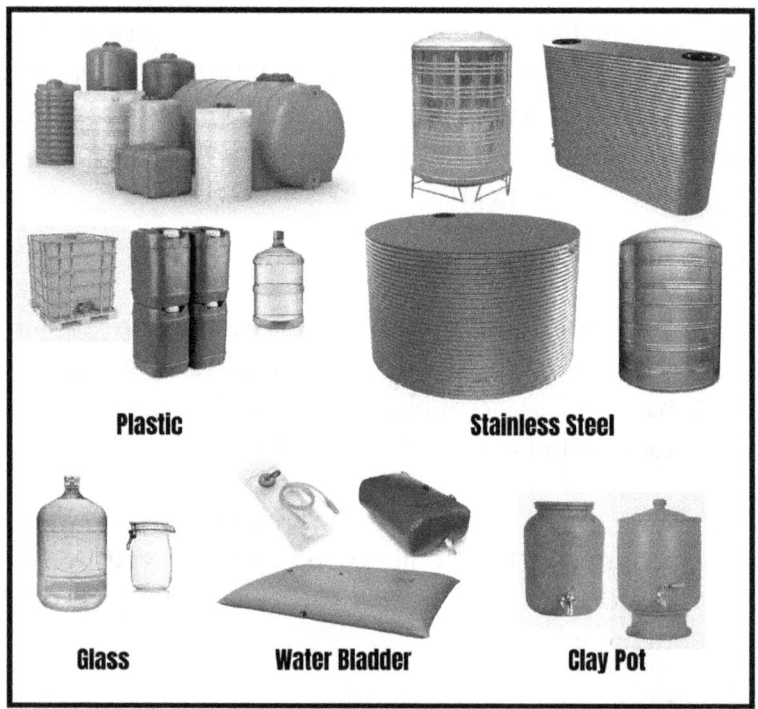

Water Storage Containers

A Closer Look at Water Tanks

Water tanks come in various shapes and sizes, catering to different needs and preferences. Let's break down the types based on where they are used and the materials they are made of.

Based on Location

Underground Water Storage Tanks:

- *Location*: As the name suggests, these tanks are buried underground.

- *Benefits*: They are hidden from view, saving space and maintaining aesthetics. Underground placement also helps with temperature regulation.

Above-Ground Water Storage Tanks:

- *Location*: Positioned at ground level or slightly elevated.

- *Benefits*: Easier access for maintenance and cleaning. They are visible and can be integrated into landscaping.

Elevated Water Storage Tanks:

- *Location*: Placed above ground level, often elevated on stands or structures.

- *Benefits*: Gravity helps with water pressure, making them suitable for supplying water to buildings. They are visible but can be incorporated into the design of structures.

Based on Materials

Concrete Water Storage Tanks:

- *Material*: Constructed from reinforced concrete.

- *Benefits*: Known for durability and resistance to fire. Suitable for large-capacity tanks.

Stainless Steel Water Storage Tanks:

- *Material*: Made from corrosion-resistant stainless steel.

- *Benefits*: Hygienic, non-corrosive, and has excellent longevity. Ideal for locations where water purity is a priority.

Fiberglass Water Storage Tanks:

- *Material*: Crafted from fiberglass-reinforced plastic.

- *Benefits*: Lightweight, corrosion-resistant, and suitable for underground or above-ground installations. Resistant to rust and decay.

Polyethylene Water Storage Tanks:

- *Material*: Made from polyethylene, a type of plastic.

- *Benefits*: Lightweight, easy to install, and cost-effective. Resistant to rust and corrosion, making them suitable for various environments.

Bolted Steel Water Storage Tanks:

- *Material*: Assembled from steel panels bolted together on-site.

- *Benefits*: Versatile and customizable for various capacities. Suitable for industrial and commercial applications. Easy to transport and assemble.

Understanding these types helps you choose the right water tank for your specific needs. Whether you prioritize aesthetics, durability, or functionality, there's a water tank type that fits your requirements.

Choosing Water Tanks

Choosing the right water tank is like finding the perfect puzzle piece for your specific needs. Here are some tips to guide you through the decision-making process:

Capacity

- *Assess your water needs*: Start by understanding your water usage. How much water does your household or facility consume regularly? Consider future needs, emergencies, or any planned expansions.

- *Select appropriate capacity*: Choose a tank that comfortably accommodates your average water usage. Having a bit of extra capacity is beneficial for unexpected scenarios.

Structure

- **Consider the location**: Where will the tank be placed? If space is limited, an overhead tank on a stand might be a good fit. Underground tanks are hidden but require specific installation.

- *Evaluate aesthetics*: If visibility matters, consider the visual impact of the tank. Some tanks can be integrated into landscaping or designed to complement the architecture.

Durability

- *Evaluate material options*: Different materials offer varying levels of durability. Consider factors like resistance to corrosion, rust, and degradation over time. For example, stainless steel and fiberglass tanks are known for their durability.

- *Think long-term*: Investing in a durable tank might involve a higher upfront cost but can save money in the long run due to reduced maintenance and longer lifespan.

Layering

- *Understand layering technology:* Some tanks come with multiple layers for added strength and insulation. For example, tanks with an inner and outer layer can resist external elements and maintain water temperature.

- *Consider climate conditions:* If you're in an area with extreme weather conditions, layering can be a crucial factor. It helps protect the tank from the impact of the environment.

Additional Considerations

- *Accessibility for maintenance:* Choose a tank design that allows for easy access for cleaning and maintenance. Regular upkeep ensures the longevity and cleanliness of the stored water.

- *Compliance with regulations*: Check if the chosen tank complies with local regulations and standards. This is especially important for commercial or industrial applications.

Installation

- *Professional installation:* Some tanks may require professional installation, especially underground or complex systems. Factor in installation costs and choose a tank that aligns with your installation preferences and capabilities.

- *Ease of installation:* If you're a DIY enthusiast, consider tanks that are relatively easy to install. Above-ground tanks, especially those designed for residential use, are often more straightforward.

Color

- *Aesthetics matter*: The color of your tank might seem like a minor consideration, but it can have an impact on water temperature. Darker colors absorb heat more readily than lighter colors.

Brand Quality

- *Research brands:* Invest time in researching and comparing different brands. Look for reputable manufacturers with track records of producing high-quality tanks. Reviews and testimonials from other users can provide valuable insights.

Water Quality

- *Material impact:* The material of the tank can influence water quality. For example, stainless steel and polyethylene are known for their inert properties, meaning they won't leach harmful substances into the water over time.

- *Consider filtration needs*: Depending on your intended use, you might need additional filtration systems. Some tanks come with built-in features to enhance water quality.

Price Range

- *Set a budget:* Determine your budget for the water tank. While it's tempting to choose the cheapest option, consider it a long-term investment. Balancing quality, durability, and cost ensures you get the best value for your money.

- *Explore financing options:* For larger tanks or commercial applications, explore financing options or grants that might be available. Some regions offer incentives for eco-friendly or water-saving installations.

Additional Considerations

- *Warranty and after-sales service:* Check the warranty offered by the manufacturer. A solid warranty indicates the brand's confidence in its product. Additionally, inquire about after-sales service and support.

- *Local regulations:* Be aware of any local regulations regarding water tanks. Some areas have specific guidelines on tank installation, and non-compliance may lead to issues in the future.

By delving into these additional considerations, you refine your decision-making process, ensuring that the water tank you choose not only meets your basic needs but also aligns with your preferences, budget, and the unique conditions of your environment.

Calculating Your Water Tank Capacity

The water tank capacity calculation involves a straightforward formula. Here's a simple guide to help you determine the ideal capacity for your needs:

- *Determine your daily water usage:* Start by estimating your daily water consumption. Consider activities like drinking, cooking, bathing, and any other water-related needs.

- *Decide on storage days:* Determine how many days you want your water supply to last in case of emergencies or disruptions. This could be a few days, a week, or more, depending on your preferences and potential scenarios.

- *Plug values into the formula:* Substitute these values into the formula. For example, if your daily water usage is 50 gallons, and you want a week's worth of water storage, the calculation would be:

Tank Capacity = 50 gallons/day × 7 days = 350 gallons

Remember, this is a basic calculation. If you have specific needs or anticipate changes in usage, adjust the formula accordingly.

Estimated Cost of Functioning Water Tank

The cost of a water tank can vary based on several factors:

Tank Type and Material:

- *Plastic tanks:* Smaller plastic tanks suitable for residential use can cost anywhere from $200 to $1,000 or more. Larger tanks or those with additional features will be more expensive.

- *Stainless steel tanks:* Stainless steel tanks are durable but can be pricier, ranging from $500 to $3,000 or more, depending on size and features.

- *Fiberglass tanks:* Fiberglass tanks typically fall in the range of $500 to $2,000, depending on size and specifications.

Installation Costs:

Installation costs vary based on the complexity of the setup. For example, underground tanks may require more labor and equipment, resulting in higher installation costs.

Additional Features:

Tanks with additional features such as filtration systems, multiple layers, or intelligent technology may come at a higher cost.

Local Factors:

Prices can also be influenced by location, materials availability, and local demand.

Remember, the total cost of having a functional water tank depends on your specific requirements and desired features. You should obtain quotes from reputable suppliers and consider the long-term benefits and savings associated with a reliable water storage solution.

Tips for Safe and Long-Lasting Water Storage

Keeping your stored water safe from contamination and ensuring its longevity involves a few simple yet crucial practices. Here are some easy-to-follow tips to maintain the purity and freshness of your stored water:

- *Container selection:* Choose containers made of

food-grade materials such as high-density polyethylene (HDPE), stainless steel, or glass. These materials are less likely to introduce harmful substances into your water.

- *Thorough cleaning routine:* Give your containers a good wash with mild soap and water before storing water. Ensure a thorough rinse to eliminate any soap residue. Periodically sanitize your containers to prevent the growth of bacteria.

- *Seal the deal:* Tightly seal your water storage containers with airtight lids or caps. This prevents conta*minants from entering and safeguards the cleanliness of your water.*

- *Location matters:* Be mindful of where you place your water storage. Keep it away from potential sources of contamination, such as chemicals or cleaning agents. Choose a secure location away from pollutants.

- *Elevate above ground:* If possible, elevate your water storage containers above ground level. This prevents contamination from flooding or runoff water, maintaining the purity of your stored water.

- *Regular rotation:* Establish a routine for using and replacing stored water. Regularly rotating your water supply ensures *it stays fresh, reducing the likelihood of contaminants settling over time.*

- *Label and date:* Label your water containers with the date of storage. This helps you track rotation, ensuring you use the oldest water first to reduce the risk of water stagnation and contamination.

- *Check for leaks:* Regularly inspect your water storage containers for leaks or damage. Promptly fix any issues to prevent contaminants from entering and maintain the quality of your stored water.

- *Monitor temperature:* Store your water containers in a cool, dark place. Elevated temperatures can accelerate the breakdown of container materials and compromise water quality.

- *Consider filtration: Integrate a water filtration system into your storage plan. This adds an extra layer of protection by removing impurities and ensuring continued purity.*

By incorporating these simple yet effective tips into your water storage routine, you can enjoy clean and reliable water for an extended period of time. Remember, a little attention goes a long way in preserving the quality of your stored water.

Space-Efficient Water Storage

When space is a precious commodity, making the most of every nook and cranny becomes a top priority. Here are some straightforward solutions for maximizing water storage in limited spaces:

- *Collapsible water storage:* Excellent choice for water storage during emergencies. They are compact, light, and reusable, and come in sizes ranging from 1 to 5 gallons. They are made of sturdy, food-safe plastic and can be easily stored when not in use. They are also easy to carry and transport, making them a practical option for anyone on the go.

- *Stackable containers:* Great option for limited storage space. They are 3-7 gallons with handles for easy movement and waste very little storage space. You can stack them anywhere and store a lot of water while using the least amount of space. Store them off the floor to prevent a chemical reaction with cement.

- *Underground water tanks:* These are durable and can store a lot of water, but the savings in above ground storage may be offset by installation costs.

- *Water Bladder:* A practical choice for storing water in case of an emergency. It requires minimal storage space when empty, but can be quickly filled up when needed. There are various types of water bladders available on the market. Some can be easily filled and stored in the bathtub, while others are heavy-duty and meant for long-term storage. There are also some options in between, which are durable but not heavy-duty, and not intended just meant to stay in the bathtub.

Incorporating these space-efficient storage options into your living space allows you to maximize your water storage without sacrificing your living area.

Interactive Element: Long-Term Water Storage Plan

Imagine having a reliable water source even in unexpected situations or emergencies. That's where a long-term water storage plan comes into play, offering a safety net for you and your loved ones

Creating a personalized long-term water storage plan involves tailoring the strategy to your needs and resources. Follow these simple guidelines to create your own plan:

- *Assess your needs:* Estimate daily water needs for each member of your household.

- *Evaluate your resources:* Identify potential sources of water (e.g., municipal supply, wells, natural sources) and storage options (e.g., tanks, barrels, bottles).

- *Calculate storage capacity:* Calculate the total amount of water storage needed based on the estimated daily needs and the duration you're planning for (e.g., 3 days, 2 weeks, etc.).

- *Choose storage solutions:* Decide on the types of containers and storage methods to use. Consider a combination of larger storage (water tanks) and smaller containers to optimize space and purification methods.

- *Plan for purification and maintenance:* Plan methods for purifying and maintaining stored water.

- *Implement your plan:* Begin collecting and storing water according to your plan.

- *Regular review and adaptation:* Set a schedule to review and update the water storage plan.

A long-term water storage plan is a wise and proactive way of protecting one of the most essential resources for our survival. It provides a sense of security, knowing that you have a dependable and customized plan to ensure a constant water supply, even in challenging circumstances.

Wrapping it Up...

In this chapter, we delved into the critical importance of having a long-term water storage plan, emphasizing its role as a reliable lifeline during emergencies. We explored how a personalized plan tailored to your needs and resources ensures a continuous and accessible water supply, offering peace of mind in unpredictable situations.

We compared various water storage containers to help you create your personalized plan, considering cost, effectiveness, and space efficiency. Whether it's the durability of stainless steel, the versatility of plastic drums, or the aesthetic appeal of glass bottles, each option has unique advantages. This comparison aims to give you insights into selecting the container that best suits your preferences and circumstances.

As you explore these container options, keep in mind your budget, available space, and the specific needs of your household. By understanding each container type's cost-effectiveness and space efficiency, you can make an informed decision that suits your lifestyle.

The next chapter will get into a crucial aspect of water safety – waterborne diseases and sanitation practices. We'll explore the potential risks associated with water contamination and discuss practical measures to ensure the cleanliness and safety of your stored water.

5

PREVENT WATERBORNE DISEASES

According to a report by the Centers for Disease Control and Prevention, 7.2 million Americans fall prey to waterborne diseases annually. These diseases, which can go unnoticed, caused 7,000 deaths, 120,000 hospitalizations, and 7 million illnesses, amounting to a staggering $3 billion in healthcare expenses (2020).

Although living away from the noise and pollution of the city seems like a great way to avoid waterborne diseases, this is not true. People who choose to live an off-grid lifestyle are actually at a higher risk of contracting such illnesses. The peacefulness of living in the wild hides the paradoxical fact that individuals embracing a self-sufficient existence may face more health hazards.

Limited access to clean and treated water, a basic necessity often taken for granted in urban settings, can become a formidable challenge, exposing the vulnerability of even the most self-sufficient individuals to the relentless tide of waterborne dangers.

As we explore further into this chapter, we uncover the complexities of living off-grid and the harsh reality that waterborne risks persist, creeping and undeterred, even in the most remote corners of our wilderness.

Common Waterborne Diseases

Water is not just a simple liquid; it is an entire ecosystem filled with various life forms, both visible and invisible. Unfortunately, some unseen creatures can be dangerous to our health and are known as waterborne diseases. These diseases are caused by microorganisms such as bacteria, viruses, and parasites that contaminate our water sources and make their way into our bodies.

Drinking contaminated water can lead to a range of health issues, from minor stomach upsets to more severe and life-threatening conditions. This is the price we pay when the water we rely on becomes a breeding ground for harmful pathogens (Most Common, 2019).

Now, rewind to Chapters 3 and 4, where we discussed purification practices and safe water storage. These were your armor against the invisible adversaries. But what if, despite your best efforts, you faltered in these crucial steps? What if the purification practices were not as foolproof as they seemed, or the water storage overlooked potential hazards?

Failing in these fundamental practices may invite waterborne diseases into the water you rely on for sustenance. Let's learn what these illnesses are so we can equip ourselves with the knowledge on how to prevent them.

Typhoid Fever

Typhoid Fever is a febrile illness caused by the bacterium Salmonella Typhi. It can cause various symptoms, from mild discomfort to severe complications. Unlike some illnesses, typhoid is a disease that affects only humans and spreads through contaminated food and water by infected individuals (Bhandari et al., 2022).

Causes

Salmonella Typhi enters your body mainly through contaminated food or water. You may consume untreated water or food prepared in unhygienic conditions, and that's all it takes for this bacterium to enter your body. Once inside, it settles in your intestines, causing chaos and typhoid fever.

This bacterium is challenging because it can survive outside the human body for a long time. This makes it harder to eradicate and highlights the importance of following strict hygiene practices to prevent its spread.

Signs and Symptoms

Typhoid fever is a stealthy foe, often starting with a gradual onset of symptoms that can be easily mistaken for other common illnesses. Early signs include a sustained high fever, headaches, and a general feeling of malaise. As the disease progresses, more distinct symptoms emerge, such as abdominal pain, weakness, and a characteristic rash called "rose spots." In severe cases, complications can arise, affecting multiple organs and posing serious health risks.

Prevention and Treatment

Preventing typhoid fever involves two critical principles: sanitation and vaccination. The paramount concern is to ensure that the water and food you consume are free from contamination. You should practice safe food handling and treat water using reliable purification methods. Vaccination is also a potent shield against typhoid, especially for those who live in or travel to areas where the disease is prevalent..

Prompt medical attention is crucial if you contract typhoid fever. Antibiotics are the primary line of defense, eliminating the Salmonella Typhi bacteria from the body. Adequate rest, hydration, and a nutritious diet support the body's recovery.

Cholera

Cholera is caused by the bacterium Vibrio cholerae and is known to be one of the most problematic waterborne diseases. This microscopic bacterium usually infects the intestines and leads to a specific type of diarrhea that can quickly become life-threatening. It is a highly contagious disease, especially in areas with poor sanitation and limited access to clean water (Ojeda Rodriguez & Kahwaji, 2020).

Causes

Vibrio cholerae is a cunning infiltrator that travels through contaminated food or water. Its primary mode of transmission is through ingestion. Imagine sipping water from a dirty source or biting into food contaminated with unclean water. These situations allow the bacteria um to invade your body and cause havoc.

Cholera thrives in unsanitary conditions. Places with poor sanitation, cramped living spaces, and limited access to clean water become breeding grounds for the bacterium, increasing the risk of outbreaks.

Signs and Symptoms

Cholera is characterized by a sudden onset of profuse, watery diarrhea, often described as "rice-water stool" due to its appearance. This rapid loss of fluids can lead to dehydration, causing symptoms such as extreme thirst, dry mucous membranes, and a rapid heartbeat. Severe cases may progress to vomiting and muscle cramps.

Prevention and Treatment

Preventing cholera depends on ensuring access to clean water and practicing proper sanitation. Good hygiene habits, such as handwashing and safe food handling, are crucial in interrupting the transmission chain. In addition, vaccines are available to provide protection against cholera, especially for those traveling to or residing in high-risk areas.

In case of infection, taking swift action is essential. Rehydration is the cornerstone of treatment, wherein oral rehydration solutions efficiently replenish lost fluids and electrolytes. In severe cases, intravenous fluids may be necessary. Antibiotics can help shorten the duration of symptoms and reduce the severity of the illness.

Giardia

Giardia, a microscopic troublemaker, is a parasite that causes an infection known as giardiasis. This sneaky intruder takes up residence in the small intestine, disrupting the normal absorption of nutrients and causing gastrointestinal distress. Giardiasis is a common waterborne illness, often spreading through contaminated water sources (Mayo Clinic, 2018).

Causes

Giardia (commonly known as "beaver fever") is a tiny parasite that can quickly enter the human body. It spreads primarily through cysts, which are protective capsules the parasite forms to help it survive outside a host. These cysts are usually found in contaminated water sources such as rivers, lakes, or untreated water supplies. Giardia can also spread through person-to-person contact in unhygienic settings.

Giardia is a tough parasite that can survive for a long time in the environment, making it a persistent threat in areas with substandard sanitation and water treatment practices.

Signs and Symptoms

The onset of giardiasis can bring about a variety of unwelcome guests in your digestive system. Symptoms often include diarrhea, abdominal cramps, bloating, and nausea. While some individuals may not show any signs of infection, others may experience more pronounced and persistent symptoms. In severe cases, giardiasis can lead to weight loss and dehydration.

Prevention and Treatment

Preventing Giardia requires a combination of cautious habits and vigilance towards water sources. You can eliminate Giardia cysts by boiling, filtering, or using water purification tablets, which ensures that the water you consume is free from these microscopic pests. Practicing good hygiene, particularly frequent handwashing, is crucial in preventing person-to-person transmission.

When giardiasis occurs, healthcare professionals prescribe antimicrobial drugs that effectively treat the infection. Staying hydrated is essential to managing symptoms, particularly diarrhea, and providing the body with the necessary support to recover.

Dysentery

Dysentery is not your run-of-the-mill stomach upset; it's a more serious intestinal condition marked by colon inflammation. This leads to severe diarrhea containing blood and mucus, accompanied by abdominal pain and discomfort. Dysentery is often caused by bacterial or parasitic infections, making it a significant concern in areas with compromised sanitation and limited access to clean water (Marie & Petri Jr., 2013).

Causes

Dysentery is a disease usually caused by specific bacteria, such as Shigella, or parasites like Entamoeba histolytica. These harmful microorganisms enter the body through contaminated food and water. For example, if unclean water is used to wash food or prepare meals, it can create a perfect entry point for these intestinal invaders.

Poor hygiene practices and crowded living conditions can worsen the spread of dysentery, making it a community-wide concern during outbreaks.

Signs and Symptoms

Dysentery announces its presence with distinctive and discomforting signs. The hallmark is bloody diarrhea, which differentiates it from other digestive issues. Alongside this, expect abdominal cramps, a constant urge to empty your bowels, and a general feeling of being unwell. In severe cases, dysentery can lead to dehydration, which can have serious consequences if not addressed promptly.

Prevention and Treatment

Preventing dysentery involves a two-pronged approach: keeping the bacteria at bay and providing timely intervention. Safe food and water practices, including thorough cooking and proper water purification, form the frontline defense. Handwashing, especially before meals, becomes a crucial habit to prevent the spread of bacteria and parasites.

If dysentery makes an unwelcome entrance, seeking medical attention is crucial. Antibiotics are often prescribed to target the specific bacteria causing the infection. Meanwhile, staying hydrated through rehydration solutions is essential to counteract the fluid loss caused by persistent diarrhea.

Escherichia Coli

Escherichia coli, or E. coli for short, is a type of bacteria that usually resides in your intestines, playing a helpful role in digestion. However, some strains of E. coli can turn rogue and cause infections. While many E. coli infections are harmless, certain strains can lead to unpleasant symptoms and, in severe cases, pose serious health risks (American Society for Microbiology, 2011).

Causes

E. coli infections often occur through the consumption of contaminated food or water. Picture a scenario where undercooked meat, unpasteurized dairy products, or contaminated fruits and vegetables make their way onto your plate. E. coli can also spread through contact with infected animals or through person-to-person transmission in settings with poor hygiene.

Certain strains of E. coli, such as E. coli O157:H7, are particularly notorious for causing illness and are often associated with foodborne outbreaks.

Signs and Symptoms

When E. coli decides to cause trouble, it announces its presence with digestive distress signals. Symptoms may include stomach cramps, diarrhea (which can sometimes be bloody), nausea, and vomiting. In more severe cases, especially in vulnerable populations like the very young or elderly, it can lead to kidney problems and other complications.

Prevention and Treatment

Preventing E. coli infections involves a combination of safe practices and vigilance. Key measures include thoroughly cooking meat, avoiding unpasteurized dairy products, and practicing proper food handling and hygiene. Ensuring that water sources are safe and well-maintained is also crucial in preventing waterborne strains of E. coli.

Staying well-hydrated is essential to managing symptoms in case of infection. Most E. coli infections resolve on their own with supportive care. Still, medical attention is necessary in severe cases or when complications arise. Antibiotics are generally not recommended for uncomplicated E. coli infections, as they may increase the risk of complications.

Hepatitis A

Hepatitis A is a viral infection that targets the liver, causing inflammation and affecting its functioning ability. While it can make you feel quite unwell, the good news is that it's typically not a chronic condition. Hepatitis A is often spread through ingesting contaminated food or water, making it a concern in areas with poor sanitation (Bintsis, 2017).

Causes

The Hepatitis A virus is contagious and can spread fast when an infected person handles food without proper hygiene. Imagine a situation where fruits, vegetables, or shellfish come into contact with contaminated water during cultivation or harvesting. In environments where sanitation is poor, the virus can quickly spread throughout the community.

If you are traveling to regions with high Hepatitis A rates, you may be at an increased risk of contracting the virus and should consider getting vaccinated.

Signs and Symptoms

Hepatitis A often announces its presence with flu-like symptoms. You might experience fatigue, nausea, and a general feeling of being unwell. As the infection progresses, characteristic symptoms such as dark urine, yellowing of the skin and eyes (jaundice), and abdominal pain may surface. The severity of symptoms can vary, and some individuals, especially children, may not show any signs.

Prevention and Treatment

Preventing Hepatitis A involves a twofold approach: vaccination and diligent hygiene practices. Vaccination is a powerful shield against the virus, providing long-lasting protection. Practicing good hand hygiene, especially after using the bathroom and before handling food, is crucial in preventing the spread of the virus.

For patients with Hepatitis A infections, there's no specific treatment to target the virus directly. However, supportive care plays a vital role. Adequate rest, staying well-hydrated, and avoiding substances that can stress the liver contribute to the body's natural healing process.

Salmonella

Salmonella can cause an infection known as salmonellosis. This sneaky intruder can enter our bodies through contaminated food, especially eggs, poultry, other meats, unpasteurized milk, and other dairy products (Giannella, 1996).

Causes

The most common way Salmonella gets into our meals is when raw or undercooked food, like chicken or eggs, is not correctly handled or cooked. These foods come into contact with other surfaces, like cutting boards or utensils, spreading the bacteria. Salmonella can also be found in the environment, making unwashed fruits and veggies potential carriers.

Poor hygiene practices during food preparation and inadequate cooking can open the door for Salmonella to enter our digestive system.

Signs and Symptoms

Salmonella doesn't go unnoticed when it stirs trouble in your digestive system. Symptoms of salmonellosis often include stomach cramps, diarrhea, nausea, and sometimes even fever. It's like your stomach's way of telling you that something isn't right. While most people recover without specific treatment, severe cases can lead to complications and might need medical attention.

Prevention and Treatment

Salmonella prevention requires easy yet crucial steps. Always cook your food thoroughly, especially poultry and eggs. During food preparation, separate raw meat from other foods to avoid cross-contamination. Also, regularly wash your hands and scrub your fruits and vegetables.

If you contract Salmonella, staying hydrated by drinking plenty of fluids to replace what you lose due to diarrhea is essential. For vulnerable populations, such as the elderly or the very young, medical attention may be required. Antibiotics may not always be prescribed, as they may not speed up recovery and can sometimes prolong the shedding of the bacteria.

Guidelines for Safe Water Consumption

Water is essential for your body, and you must ensure that your water is safe and reliable. Choose your water sources wisely. Whenever possible, opt for treated or purified water. If you are unsure about the source of the water, consider boiling it for at least one minute to eliminate harmful microorganisms.

Secondly, maintain good hygiene. Always wash your hands before handling water or utensils used for drinking. Use clean and sanitized containers for water storage, and avoid touching the inside of the container or the rim where you drink from. These simple steps will help you ensure your water is safe to drink and will help keep you healthy.

Checking, Treating, and Storing Water During Outbreaks

During disease outbreaks, your water vigilance should kick into high gear. Regularly check your water source for signs of contamination, like discoloration or unusual odors. If you're in doubt, it's time for water treatment. Boiling is a reliable method, but consider using water purification tablets or filters if that's not feasible. Be diligent in following the instructions on these treatments to ensure their effectiveness.

Storage matters, too. Use covered containers to protect your water from potential contaminants. If possible, keep your water supply in a cool, dark place to prevent the growth of harmful microorganisms.

Contaminated Water Fallout

Let's discuss how to handle the situation if someone in your group unintentionally drinks contaminated water. Don't worry; stay calm, and watch out for any signs of waterborne illnesses such as diarrhea or nausea. Encourage the person to drink safe water and seek medical attention if their symptoms persist or worsen.

Remember the importance of water purification that we discussed in Chapter 3. Those methods are not just suggestions, but they are your protection against the hidden dangers that can be present in untreated water. Purification is the first line of defense, preventing waterborne diseases from entering your life. Whether you are at home or facing the challenges of an outbreak, let the lessons from our purification chapter guide you toward maintaining a clear and safe path for the water you consume. Stay alert, stay informed, and let every sip be a celebration of your health and well-being.

Waterborne Diseases Summary Table

Disease	Causes	Signs and Symptoms	Prevention and Treatment
Typhoid Fever	Bacterial infection (Salmonella Typhi) from contaminated food and water	Sustained high fever, abdominal pain, weakness	Vaccination, safe food and water practices, antibiotics for treatment
Cholera	Infection with Vibrio cholerae bacteria through contaminated water or food	Profuse watery diarrhea, dehydration, vomiting	Hygiene, safe water and food practices, vaccination in high-risk areas, rehydration for treatment
Giardia	Parasitic infection (Giardia lamblia) from contaminated water or food	Diarrhea, abdominal cramps, bloating	Boiling, filtering, purification tablets for water, good hygiene, medications for treatment
Dysentery	Bacterial or parasitic infection (e.g., Shigella) through contaminated food and water	Bloody diarrhea, abdominal cramps, dehydration	Safe food and water practices, hygiene, antibiotics for bacterial dysentery, supportive care for treatment
Escherichia Coli	Infection with E. coli bacteria from contaminated food, especially undercooked meat	Stomach cramps, diarrhea, nausea	Cooking meat thoroughly, hand hygiene, avoiding unpasteurized products, rehydration for treatment
Hepatitis A	Viral infection (Hepatitis A virus) from contaminated food or water	Flu-like symptoms, jaundice, dark urine	Vaccination, hand hygiene, safe food and water practices, supportive care for treatment
Salmonella	Bacterial infection (Salmonella) from contaminated food, especially poultry and eggs	Stomach cramps, diarrhea, nausea	Cooking food thoroughly, hand hygiene, avoiding cross-contamination, supportive care for treatment

Wrapping it Up...

In this chapter, we have explored the complex landscape of waterborne diseases. We have uncovered their causes, signs, and ways to prevent them. To navigate potential health risks, it's essential to understand and be prepared for and understand these diseases. Our best defense is to adopt safe water consumption practices, such as selecting clean water sources and practicing vigilant hygiene. These measures are essential to safeguard against waterborne threats.

It is important to understand that the above practices are helpful in survival situations and play a vital role in maintaining a healthy daily life. By implementing these practices, we can ensure that our water promotes our well-being instead of causing health concerns. Looking towards the future, let's explore water conservation and recycling techniques that can help us achieve sustainable water practices. This journey will not only positively impact the environment but also contribute to your health.

6

INNOVATIVE WATER CONSERVATION PRACTICES

A crisis is looming worldwide due to the intersection of two critical elements: - energy and water. A startling revelation has emerged as researchers investigate the intricacies of our ever-increasing energy demands. It is predicted that by the year 2040, a worldwide water shortage will occur, not due to the usual suspects of climate change or population growth but as a direct consequence of our insatiable appetite for energy (Prupis, 2014).

The connection between energy and water is becoming increasingly apparent in today's world, and the two are inextricably linked. As our demand for energy rises, so too does our water usage, setting the stage for a future crisis of unprecedented proportions.

The facts and projections surrounding energy and water consumption reveal a concerning trend. We are consuming more water to satisfy our growing energy needs. This reality has implications extending far beyond the power plants and water sources we rely on. It highlights the urgent need for water conservation, which has become a pressing concern rather than a mere environmental consideration.

This raises a fundamental question: Can humanity redefine its relationship with energy and water before it's too late? The answers to this question are yet to be discovered and are intricately woven into the complex fabric of our world and its many challenges. The task at hand is to find a way to balance our energy-dependent civilization with the imperative to protect our planet's most precious resource: water.

The Water Footprint

If you're reading this, it's clear that you're serious about water as a valuable resource. As you learn more about this topic, you'll come across the term "water footprint," which is an important measure to understand. Simply put, a water footprint represents the total amount of freshwater used by an individual, community, or business entity, directly and indirectly. This includes the water we use at home or at work, in addition to and the water used to produce the goods and services we consume. It's a comprehensive measure that encompasses our actions and the complex supply chains that support our modern lifestyles.

Consider your morning routine: the shower that wakes you, the coffee that kick starts your day, the clothes you wear, and the food you consume—all have a water footprint. Beyond the obvious uses, like drinking and bathing, our daily choices—what we eat, what we wear, and what we buy—have far-reaching implications for water resources.

For instance, that cup of coffee you enjoy daily carries a hidden water cost. Water is essential, from growing the coffee beans to brewing your cup. The clothes you wear, the gadgets you use, the meals you prepare—all leave an indelible mark on your water footprint.

Let's connect this idea to the main focus of our chapter—the growing water crisis due to our increasing energy consumption. Your water footprint is more than just a theoretical concept. Still, it's a physical representation of the choices you make every day. The water used to produce electricity, the water needed for manufacturing, and the water in the products you buy all contribute to your personal water footprint.

As we face the prospect of a global water shortage by 2040, we must understand the impact of every action we take on water availability. From the food we buy to the energy we consume, everything affects the intricate web of water resources. It is not just our responsibility to be mindful of our water consumption but also necessary to safeguard our planet's most valuable resource.

We must reflect on our water footprint and realize the significance of each decision that we make in shaping our water-scarce future. Every drop of water we save today contributes to a more sustainable and resilient tomorrow.

Efficient Water Use

To minimize our impact on the planet and reduce our water usage, it's essential to adopt simple yet effective practices for efficient water use at home. It's not just a choice we make; it's our responsibility. Let's look at some practical guidelines that can make a significant difference:

- *Fix those leaks:* Leaky faucets and dripping pipes may seem inconspicuous, but they contribute to significant water loss over time. Regularly check and promptly fix any leaks around the house to prevent wastage.

- *Mindful showers and baths:* Long, luxurious showers may be tempting but consume substantial water. Consider shortening your shower time and installing water-efficient showerheads. Also, opt for a shower over a bath when possible, as baths use more water.

- *Water-efficient appliances:* Upgrade to water-efficient appliances, especially washing machines and dishwashers. Look for models with high energy and water efficiency ratings to reduce consumption while maintaining functionality.

- *Collect rainwater:* Harness the power of nature by installing rain barrels to collect rainwater. This harvested water can be used for watering plants or even cleaning outdoor spaces, reducing reliance on treated water for such tasks.

- *Smart lawn care:* If you have a garden or lawn, water it in the early morning or late evening to minimize evaporation. Consider using a soaker hose or drip irrigation system to deliver water directly to the roots, reducing waste.

- *Reconsider flushing habits:* Every flush uses a significant amount of water. Consider installing a low-flow toilet or placing a filled bottle in the tank to displace water, reducing the amount used per flush.

- *Opt for full loads:* Wait until your dishwasher or washing machine is full before running them. This not only conserves water but also maximizes the efficiency of each cycle.

As you follow these guidelines, it is crucial to understand how your actions impact your water footprint. The water you use for everyday activities, such as washing dishes or doing laundry, contributes to this measure, which is unseen but significant. By adopting water-efficient practices at home, you can actively reduce your personal water footprint.

Think of it as a ripple effect - the water saved in your home extends beyond your immediate surroundings. It contributes to a collective effort to conserve water resources globally, aligning with the broader narrative of our chapter. Each mindful choice becomes a conscious step towards a sustainable future, where our relationship with water is one of respect, conservation, and harmony.

So, as you turn off the tap while brushing your teeth or opt for a shorter shower, remember that these seemingly small actions accumulate into a powerful force for positive change. It's not just about saving water; it's about reshaping our habits to leave a gentler mark on the planet and ensure a thriving future for future generations.

The Potential of Graywater

As the world faces an impending water crisis, exploring new and innovative ways to recycle and reuse water is crucial. One such method that holds great promise is the use of graywater. Let's take a closer look at what it is and the many benefits it offers.

Graywater is the term used for domestic wastewater from washing dishes, doing laundry, bathing, and showering. Unlike blackwater, which comes from toilets and contains fecal matter, graywater is relatively clean and can be reused for non-potable purposes such as watering plants or flushing toilets. Using graywater can bring us the following advantages:

Conservation of Freshwater

The One primary benefit of using graywater is the conservation of freshwater. Instead of allowing water from showers and washing machines to flow down the drain and mix with sewage, graywater can be captured, treated, and repurposed for other household needs. This reduces the demand for fresh, potable water for non-potable activities.

Environmental Impact

The environmental benefits of graywater reuse are multifaceted. Recycling water within the household reduces reliance on extracting and treating large volumes of freshwater. This, in turn, reduces the energy and resources expended in water treatment processes, contributing to a more sustainable and eco-friendly water management approach.

Support for Plant Life

When appropriately treated, graywater can be an excellent source of plant nutrients. This recycled water, rich in biodegradable soap residues and organic matter, can enhance soil fertility and support the health of gardens and landscapes.

Mitigation of Urban Heat Island Effect

Graywater can be used for irrigation in urban environments, helping to mitigate the heat island effect. It helps cool the surroundings and foster a more pleasant and sustainable urban ecosystem when applied to green spaces.

Community-Level Impact

Widespread adoption of graywater reuse practices within communities can lead to a significant reduction in overall water demand. This collective effort, along with other water conservation measures, can significantly contribute to solving the challenges posed by water scarcity.

Graywater is a valuable resource that, when harnessed and managed responsibly, can play a crucial role in reshaping our approach to water usage. As we navigate the complex web of water conservation, embracing the potential of graywater is not just an option but a meaningful step towards a more sustainable and water-resilient future.

Tips for Everyday Use

Graywater is often underutilized. However, it holds tremendous potential for water conservation. Here are some easy ways to incorporate graywater into daily routines:

- *Collecting shower water:* While waiting for your shower water to heat up, collect the initial cold water in a bucket or basin. This water, which would otherwise go down the drain, can be used to water plants, flush toilets, or even clean.

- *Reusing laundry water:* Consider directing the graywater to a collection basin when your washing machine completes a cycle. Free from harsh detergents, this water can be repurposed for outdoor irrigation, giving your plants a nutrient boost.

- *Bathroom sink graywater:* Install a simple system to capture water from your bathroom sink. This water can be utilized for tasks such as watering indoor plants or cleaning surfaces.

- *Graywater diversion systems:* Explore graywater diversion systems that can be installed in your home plumbing. These systems redirect graywater from specific sources, like showers and washing machines, to a separate storage or distribution system for later use.

- *DIY graywater irrigation:* Create a DIY graywater irrigation system for your garden. Use perforated pipes or soaker hoses to distribute graywater directly to the root zones of plants, minimizing water wastage through evaporation.

- *Mulching with graywater:* Apply graywater directly to the base of trees and shrubs, then cover the area with mulch. This conserves water by reducing evaporation and enhancing the soil's ability to retain moisture.

- *Graywater-compatible detergents:* For household use, choose biodegradable and graywater-friendly detergents. These products break down more quickly, ensuring that the graywater generated *is suitable for reuse without harming plants or the environment.*

- *Timing matters:* Be mindful of when you use graywater. Watering plants in the early morning or late afternoon reduces evaporation losses, maximizing the effectiveness of graywater use.

- *Educate and involve others:* Spread awareness within your community about the benefits of graywater reuse. Encourage neighbors and friends to adopt similar practices, collectively impacting water conservation.

- *Compliance with local regulations:* Before implementing graywater reuse systems, familiarize yourself with local rules and guidelines. Some areas may have specific requirements or restrictions regarding graywater use, and compliance ensures efficiency and adherence to legal standards.

By adopting these practices, you can conserve precious water resources and take a more sustainable and mindful approach to daily living. Remember, every drop saved is a step toward a water-resilent future, and by embracing the potential of graywater, you can become an integral part of this positive change.

Water Recycling

Water recycling systems, also known as water reuse or wastewater recycling systems, are innovative approaches to managing our precious water resources. These systems are designed to treat and repurpose wastewater, once considered "used" water, turning it into a valuable and sustainable resource.

Sophisticated setups of water recycling systems treat wastewater from various sources, making it safe for reuse in non-potable applications.

Large Scale Water Recycling Systems

- *Collection and pre-treatment:* The process begins with collecting wastewater from different sources, including sewage, industrial discharges, and stormwater. Before treatment, the water undergoes a preliminary stage of screening and grit removal to eliminate large debris and particles.

- *Primary treatment:* In the primary treatment phase, physical processes like sedimentation and flotation are employed to remove suspended solids and further separate larger particles from the water.

- *Secondary treatment:* The water then moves to the secondary treatment, where biological processes take center stage. Microorganisms break down organic matter, reducing pollutants and contaminants in the water. This phase is crucial for enhancing water quality.

- *Advanced treatment:* Advanced treatment methods may be employed depending on the desired quality of the recycled water. These can include processes like membrane filtration, reverse osmosis, or UV disinfection to achieve high purity.

- *Distribution and reuse:* Once the water has undergone the necessary treatment, it is ready for reuse. Recycled water can be distributed through separate pipes or channels and utilized for various non-potable purposes. Typical applications include landscape irrigation, industrial processes, and cooling systems.

When installed and maintained properly, water recycling systems offer the following benefits:

- *Conservation of freshwater:* These systems recycle and reuse wastewater, contributing to preserving freshwater resources. This is especially crucial in regions facing water scarcity or drought.

- *Reduced environmental impact:* Treating and recycling wastewater reduces the environmental impact of discharging untreated water into rivers or oceans. It helps protect aquatic ecosystems and minimizes pollution.

- *Economic savings:* Water recycling systems offer economic benefits by reducing the demand for fresh, treated water. This, in turn, can lower water bills for municipalities, industries, and communities.

- *Mitigation of water scarcity:* Recycling systems offer a sustainable solution to meet water demands in areas grappling with water scarcity. They provide an additional and reliable source of water for various applications.

- *Community resilience:* By diversifying water sources, implementing water recycling systems enhances community resilience. This reduces dependence on a single water supply and provides a buffer against disruptions or shortages.

It's hard to deny that water recycling systems are a perfect example of a sustainable approach to water management. These systems transform wastewater into a valuable resource, making them crucial in securing a water-resilient future for communities and the planet. Although these systems are primarily designed for large communities, an off-grid graywater system can serve individual properties without needing major infrastructure.

Interactive Element: DIY Graywater System

Creating a graywater system at home is a rewarding and eco-friendly venture. It allows you to recycle water from your daily activities, reducing your environmental footprint. Here's a step-by-step guide to help you set up your own graywater system:

Materials and Tools You'll Need:

- bucket or basin
- PVC pipes or hoses
- pipe fittings
- mulch or gravel
- gloves and safety gear
- water-tight sealant
- container for storing graywater

- valves (optional for control)

- diversion device (optional for automatic redirection)

Graywater System Estimated Cost

To estimate the cost of installing a graywater system, you need to consider several factors, such as the system's complexity, whether it is a DIY project, and the specific components you choose. Here is a breakdown of the potential costs involved to give you a general idea:

DIY vs. Professional Installation:

- *DIY project:* The costs can be relatively low if you create your own graywater system. You will mainly be spending money on materials such as pipes, fittings, and a storage container. The cost will depend on the size of your project and the materials you choose, ranging from $100 to $500.

- *Professional installation:* You will likely incur additional costs when hiring a professional installer. The fees can vary depending on your location, the system's complexity, and the installer's rates. Expect to pay $500 to $2,000 or more for professional installation.

Materials:

- *PVC Pipes and Fittings:* PVC pipes are a common choice for graywater systems.

- The cost depends on the length and diameter of the pipes, ranging from $1 to $5 per foot. Fittings can add an additional $1 to $20 each.

Storage Container:

- A storage container or tank can cost from $50 for a basic barrel to a few hundred dollars for a larger, more sophisticated tank.

Optional Components:

- *Diversion device:* If you opt for an automatic diversion device, it can add around $100 to $300 to your costs. Manual valves are less expensive, but still contribute to the overall cost.

- *Filters:* Simple filters to remove larger particles might range from $10 to $50.

Mulch or Gravel:

- Creating a mulch basin can be an affordable DIY project that requires minimal effort and materials. The cost may vary depending on the amount and type of mulch or gravel used, but it typically falls within the range of $20 to $100.

Professional Consultation:

- If you seek professional advice or consultation before starting your project, this might add an additional cost, typically ranging from $50 to $200 per hour.

Permits and Fees:

- Graywater system installations may require permits that, depending on your location, cost between $50 and a few hundred dollars.

Maintenance Costs:

- While routine maintenance costs are generally low, it's a good idea to set aside a small budget for occasional replacements or repairs. This might range from $20 to $100 per year.

Remember, these estimates are general and can vary based on your specific circumstances. DIY projects offer cost savings, but they also require time and effort. Professional installations provide convenience but come with additional expenses. Regardless of your route, investing in a graywater system can contribute to water conservation and result in long-term savings on water bills.

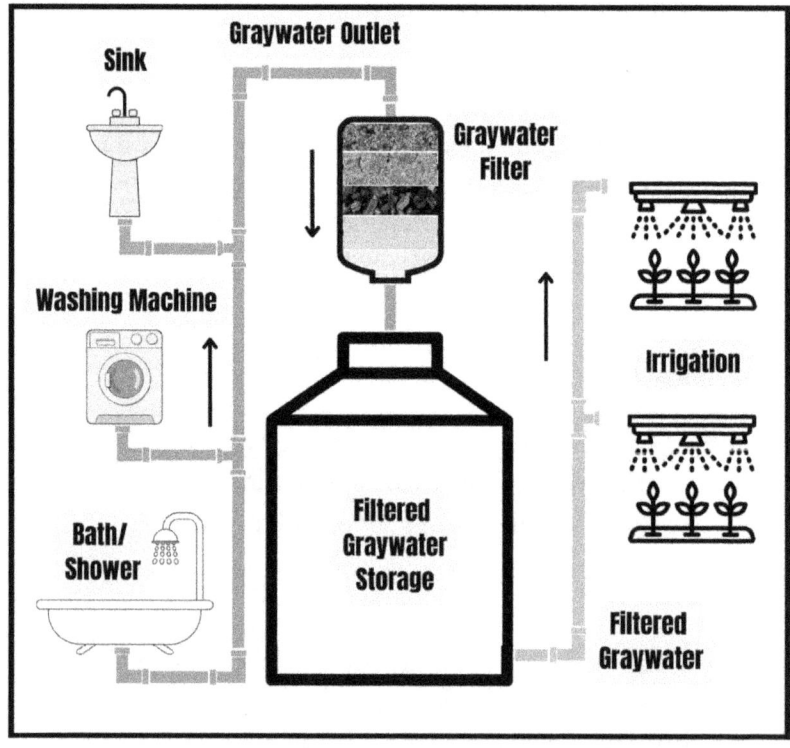

Greywater System

Creating a graywater system for your home is a sustainable and eco-friendly way to reuse water. Check out the following table for steps to make your own graywater system:

Step	Action Steps	Description
1	Assessment of Water Usage	Evaluate your household water usage to determine the potential for graywater recycling. Identify areas such as the bathroom and laundry where graywater can be easily collected.
2	Check Local Regulations	Investigate local regulations and guidelines regarding graywater systems. Ensure compliance with legal requirements or permits to install and use a graywater system.
3	Select graywater Sources	Choose specific sources of graywater, typically from showers, bathtubs, and washing machines. Avoid water from toilets or kitchen sinks, as it may contain contaminants that are harder to treat.
4	Diverting graywater	Install a diverter valve to redirect graywater from selected sources to the graywater system. Depending on your preference and budget, this can be a manual or automatic system.
5	Filtering and Treatment	Integrate a filtration system to remove debris and particles from the graywater. Consider additional treatment methods like UV sterilization or chlorine tablets to ensure water quality.
6	Storage Tank Installation	Set up a storage tank to hold the filtered and treated graywater. The tank should be appropriately sized based on water usage and local climate conditions.

Step	Action Steps	Description
7	Distribution System	Implement a distribution system to carry graywater to irrigation points. This could involve using gravity, a pump, or a combination of both to transport water to the desired areas.
8	Selecting Irrigation Methods	Choose appropriate irrigation methods for your landscape, such as drip or subsurface irrigation. Ensure that the chosen methods match the water requirements of your plants.
9	Mulching and Soil Preparation	Apply mulch around plants and prepare the soil to enhance water absorption and prevent surface runoff. This contributes to the efficient use of graywater for plant growth.
10	Regular Maintenance	Establish a routine maintenance schedule. This includes checking and cleaning filters, inspecting pipes for leaks, and ensuring the system's overall functionality.
11	Educate Household Members	Inform everyone in your household about the graywater system, including guidelines on what can and cannot be disposed of in the graywater. Promote water conservation habits among family members.
12	Monitor Water Quality	Periodically test the graywater quality to ensure that it meets safety standards. Adjust the treatment processes if necessary to maintain water quality.

Following these steps, you can create an efficient and environmentally friendly graywater system tailored to your household's needs. Remember to monitor and maintain the system for optimal performance.

Wrapping it Up...

In this chapter, we discussed the critical issue of the upcoming global water crisis. Research has projected a worldwide water shortage by 2040 due to increasing energy demands, emphasizing the need for proactive water conservation measures.

We also explored the concept of a water footprint and how our daily activities contribute to our individual impact on water usage. The significance of being mindful of our water footprint was highlighted. It reinforced the idea that every action we take directly influences this global challenge.

As you've progressed through the chapters, it has become evident that now is the time to act. Water conservation is an option and a responsibility we all share. By following the guidelines for efficient water usage at home, appreciating the importance of recycling graywater, and understanding the possibilities of water recycling systems, you have gained valuable insights into practical measures for sustainable living.

The next chapter will guide you in creating your own water supply from scratch, using the knowledge acquired from Chapters 2–6. Get ready to take control of your water future and explore the transformative potential within your reach.

7

7-Step Water Survival Plan

A recent study published on ResearchGate has revealed that having an at-home water supply system brings significant and measurable benefits compared to traditional water supply systems (Evans, 2013).

This chapter is a group effort that combines the details from Chapters 1-6 to assist you in building your off-grid water supply. Chapter 5 puts a spotlight on waterborne diseases, which are a crucial aspect of water supply systems.

The *7-Step Water Survival Plan,* based on the insights presented in previous chapters, emphasizes the need to understand the broader context of home water supply and navigates the complex waters of this important topic. Follow these steps to build resilience y and ensure a reliable home water supply.

Step 1: Your Water Sources

Let's explore the top water sources highlighted in Chapter 2—Groundwater, Rainwater, and Surface Water—and connect them with the water conservation practices from Chapter 6. We can then conduct a comprehensive water assessment, considering the insights of waterborne diseases discussed in Chapter 5.

Groundwater

Groundwater is an abundant source of clean water and an effective solution for off-grid water supply. By using resources like topographic maps, satellite imagery (such as Google Maps), and exploration, you can assess your landscape and identify the groundwater potential. Digging test wells can help you confirm your findings.

Rainwater

Rainwater is an exceptional off-grid water source. It is truly a gift from nature. By collecting these droplets, you use nature's abundance, which complements the water-saving tools introduced earlier. Making the most of every raindrop guarantees sustainability and promotes conservation.

Surface Water

Use surface water from rivers, lakes, and streams to create a sustainable off-grid water system. It is a natural reservoir that saves water and benefits the environment.

Water Source Assessment

When conducting a water assessment, it is essential to balance conservation and health. To do so, you can use the water-saving practices and tools mentioned in Chapter 6 and Bonus Chapter 2. These include using water-efficient appliances and intelligent water usage practices.

During the assessment, watch for potential contaminants and health hazards. To ensure safe and healthy water, follow the recommended guidelines for safe drinking water from Chapter 5.

Step 2: Your Water Collection

In this pivotal step, we'll explore our top contenders for collecting water. Let's seamlessly connect these choices to the water-saving tools we discovered in earlier chapters and conduct a thorough water assessment with an eye on the insights from Chapter 5.

Groundwater Collection

A water well is a great way to access underground water reserves, providing a reliable and high-quality water supply for all your needs.

Depending on water table depth and ground conditions, wells can be drilled, driven, or dug. Understanding contamination risks is also crucial, as surface runoff and proximity to septic systems can introduce pollutants. Maintaining a safe distance becomes a pivotal practice to mitigate these risks.

Rainwater Harvesting

Rainwater harvesting is a practical, eco-friendly method of tapping into a natural resource for various uses. While it offers numerous benefits, location, setup costs, and reliance on rainfall should be considered when implementing a rainwater harvesting system.

DIY Rain Barrel Water System

The rain barrel water system is a simple, cost-effective way to harvest rainfall. It uses sturdy barrels to store rainwater directly from downpipes.

Here is a do-it-yourself task to create your own rain barrel water system:

Gather your materials:

- Obtain a sturdy barrel (following the suggestions in Chapter 4's suggestions).
- Acquire a downspout diverter kit to channel rainwater into the barrel.
- Gather a few tools, such as a saw, drill, and a level.

Select a strategic location:

- Choose a spot near a downspout to maximize rainwater collection.
- Ensure the ground is level to stabilize the barrel.

Install the downspout diverter:

- Follow the kit instructions to connect the diverter to the downspout.
- Set up the diversion mechanism to direct water into the barrel.

Create an overflow system:

- Attach an overflow hose to prevent waterlogging during heavy rains.
- Position the overflow hose to direct excess water away from your foundation.

Screen the barrel opening:

- Install a fine mesh or screen over the barrel opening to filter out debris.

- This ensures the collected rainwater is free from contaminants.

Regular maintenance:

- Periodically clean the barrel and the filter to maintain water quality.
- Check for any wear or damage and repair as needed.

Integrate into your water system:

- Connect a spigot to the barrel for easy access to collected rainwater.
- Integrate the rain barrel into your overall water management system.

By crafting your rain barrel water system, you actively contribute to sustainable water practices. This DIY approach adds a personalized touch to your off-grid water survival.

Surface Water Collection

Surface water is usually pumped from its source and filtered at the intake before being stored. It can then be used for irrigation or treated for human consumption. It is important to ensure that all necessary legal permits have been obtained and that the source can provide the required flow rate for your water usage needs.

Step 3: Your Purification Strategy

Chapter 3 discussed the critical task of selecting an appropriate water purification process. As we examine the top off-grid purification options, remember that using multiple methods increases flexibility and redundancy.

Chlorination

Adding chlorine to the water will battle and neutralize microscopic invaders like bacteria and viruses. It's a robust, cost-effective, and widely used chemical purification method. It is less effective in removing chemical pollutants, heavy metals, and other non-biological impurities.

Filtration

Filtration involves passing water through physical barriers that trap impurities. It removes particles and contaminants to ensure a crystal-clear and safe water supply. This affordable and versatile method has limited effectiveness against some contaminants, such as heavy metals and chemical pollutants.

Carbon Filtration

Activated carbon acts as an absorber of toxins and impurities. It's power lies in purifying water by adsorbing a wide range of contaminants, making it an excellent off-grid choice. It is perfect for taste and odor but less effective for minerals, salts, and dissolved substances.

Reverse Osmosis

Reverse osmosis employs a semi-permeable membrane to separate impurities from water molecules. It's like creating a barrier that only allows the purest liquid to pass through. This gold standard in purification works at a molecular level.

RO is limited in its ability to remove some volatile organic compounds and gases. It also consumes a lot of energy and generates wastewater as a by-product.

Solar Disinfection

The most straightforward and cost-free purification option that only requires exposure to sunlight to kill pathogens. However, it depends on local weather conditions and does not remove chemical or physical impurities.

Purification Method Assessment

Ensure that your choices align with the principles outlined in the water purification chapter so that your method is dependable and tailored to your off-grid situation's specific challenges.

Be mindful of maintenance requirements to understand the long-term commitment of your chosen purification strategy.

Consider adopting multiple purification methods to provide flexibility and redundancy for your water survival plan.

Step 4: Your Water Storage Choice

In this fourth chapter of your off-grid water journey, we'll discuss selecting the ideal storage container for the third step of your water survival plan. Chapter 4 outlines various types and sizes, allowing you to choose the perfect fit for your off-grid circumstance.

Before starting, be sure to calculate your water storage requirements to better understand your long-term storage strategy.

Plastic Barrels

Made from high-density polyethylene (HDPE), barrels are versatile containers for storing and transporting liquids. They are a budget-friendly option in different sizes to suit your needs.

Stainless Steel Containers

Stainless steel containers are built to last. They are corrosion resistant, do not leach harmful substances, and have impressive thermal properties to maintain water temperature for extended periods.

Glass Bottles or Jars

Glass bottles are non-porous, do not absorb odors, and are fully recyclable. Their initial cost is generally higher than that of plastic, but they are durable and less prone to discoloration.

Water Bladders

Small water bladders, sometimes called hydration packs, are portable collapsible bags. Useful on the go in emergency situations. Larger-capacity bladders are an alternative to water tanks and can be stored in confined spaces.

Clay Pots

The advantage of clay pots is their porous nature, which creates a natural cooling effect through evaporation. Clay is a raw material that interacts with water, which absorbs minerals from the clay, providing some minor health benefits.

Water Tanks

Water tanks are the giants of water storage. They come in various shapes, sizes, and materials, typically plastic, fiberglass, or stainless steel.

They can be located above, at, or below ground level. Their large capacity offers long-term water security, but they have higher upfront costs.

Storage Assessment

The significance of selecting suitable containers cannot be overstated. It's not merely about holding water. It's about ensuring your efforts translate into a reliable, secure water supply. A misstep in container choice can compromise the integrity of your stored water, making all your meticulous planning seem like an exercise in futility.

Consider combining storage options, such as storing your primary water source in tanks and storing drinking water purified by your chosen method in smaller containers.

Step 5: Your Water Recycling System

In Chapter 6, we ventured into creating a water recycling system where water takes on new life, contributing to sustainability and conservation. Let's revisit the concept of a graywater system and include it in your survival plan.

Creating a graywater system transforms used water into a valuable resource, contributing to the sustainability of your off-grid water supply.

Step 6: Testing Your Water Supply System

Testing your water system is like giving your creation a health check. It ensures that the system operates effectively and troubleshoots common issues. Let's explore techniques for testing your water filtration system and solving contamination, mineral build-up, and low water pressure in off-grid systems.

Water quality testing kits:

- Invest in water quality testing kits available in hardware stores.
- These kits often include test strips or chemical solutions to detect impurities, ensuring your water meets safety standards.

Microbial analysis:

- Send water samples to a laboratory for microbial analysis.
- This thorough examination helps identify any harmful bacteria, ensuring your filtration system is effective in removing contaminants.

DIY clarity test:

- Fill a clear glass with filtered water.
- Observe any cloudiness, unusual color, or floating particles. Clear water indicates effective filtration.

Common Issues and Solutions

Contamination: Identifying and eliminating intruders

- *Issue*: Unwanted particles or microorganisms in the water.
- *Solution*: Increase the efficiency of your filtration system, consider upgrading filters, or consult water quality experts for tailored solutions.

Mineral build: softening the blow

- *Issue*: Accumulation of minerals leading to hard water.

- *Solution*: Install water softeners to reduce mineral content. Regularly clean and maintain the system to prevent build-up.

Low water pressure: Pumping life back in

- *Issue*: Insufficient water pressure from the pump.

- *Solution*: Check for clogs or debris in the pump. Ensure proper wiring and power supply. If using a solar pump, ensure the panels receive adequate sunlight. Regular maintenance and prompt repairs address low-pressure concerns.

By testing and troubleshooting your water system, you ensure its health and prolong its efficiency. So, let your water system thrive, and enjoy the peace of mind that comes with a healthy and reliable off-grid water supply.

Step 7: Maintain, Check, and Readjust Your System

Think of the seventh step of your survival plan as the ongoing care and fine-tuning needed to keep the water flowing harmoniously. To make that happen, let's examine the importance of regular inspections and explore tips for maintaining your off-grid water system for optimal, long-lasting performance.

Preventing issues before they emerge:

- Regular inspections allow you to catch potential problems early on.

- Identifying and addressing issues proactively prevents larger, more costly repairs down the line.

Ensuring efficient water flow:

- Checking the system ensures that water flows smoothly and consistently.

- Regular inspections help maintain optimal pressure, preventing low water flow issues.

Prolonging system lifespan:

- Routine maintenance and adjustments contribute to the longevity of your water system.

- It's like giving your system a regular health check, ensuring it stays robust and reliable.

Maintenance Tips

Establish a regular inspection schedule:

- Set specific intervals for system inspections (e.g., quarterly or bi-annually).

- Consistency in inspections ensures comprehensive coverage and timely identification of any issues.

Check for leaks and drips:

- Inspect pipes, connections, and fixtures for any leaks or drips.

- Promptly repair any identified issues to conserve water and prevent damage.

Monitor pump performance:

- Regularly assess the performance of your water pump.

- Listen for unusual noises, check for vibrations, and ensure the pump is delivering water at the expected pressure.

Clean filters and screens:

- Periodically clean or replace filters and screens as recommended by the manufacturer.

- This prevents clogs and ensures the filtration system operates efficiently.

Inspect solar panels (if applicable):

- If your system includes solar panels, inspect them for dirt or debris.

- Clean panels regularly to maximize energy absorption and maintain the efficiency of solar-powered components.

Educate and involve:

- Teach household members to recognize signs of potential issues.

- Involve everyone in responsible water usage practices, fostering a collective effort in maintaining the system.

Adjust based on seasonal changes:

- Be mindful of seasonal variations that may affect water availability.

- Adjust settings or practices accordingly to optimize the

system's performance under changing conditions.

By being proactive about maintenance, you care for your water system's well-being. So, watch your off-grid water system, and enjoy a reliable and constant water flow as you journey towards sustainability.

Interactive Element: Your Water Survival Plan and Goals

With the extensive insights gained from chapters 1-6, it's time for you to craft your own personalized *Water Survival Plan*. Each chapter has unveiled essential aspects of sustainable water management. Now, armed with this comprehensive understanding, the task is yours to weave these elements into a cohesive and tailored plan that perfectly aligns with your specific location, needs, and aspirations.

Your Water Survival Plan Fillable Table:

Type	Details
Daily Water Consumption (g/day)	[Input]
Water Sources • Primary • Secondary	[Input]
Water Collection • Primary • Secondary	[Input]
Water Purification • Primary • Secondary	[Input]
Water Storage • Primary • Secondary	[Input]
Water Recycling • Greywater	[Input]
Water Testing	[Input]
System Maintenance	[Input]
Additional Information	[Input]

Notes and Goals For Your Water System

Notes:

Goals:

1. [Input]

2. [Input]

3. [Input]

Feel free to print this chart and checklist or copy it to a document where you can fill in the specific details related to your location and needs. This tool should help you systematically plan, track, and achieve your goals for an effective and sustainable off-grid water supply system.

Wrapping it Up...

In this chapter, the key emphasis lies in the critical importance of meticulous planning and thorough assessment before venturing into the creation of an off-grid water system. It underscores that off-grid living demands adaptability and ongoing maintenance, acknowledging challenges while highlighting that the rewards of achieving a self-sufficient water supply far exceed these obstacles.

As you step into crafting your personalized plan, the chapter encourages a creative and resourceful approach, urging you to find inventive solutions tailored to your unique needs and environmental conditions. It sets the stage for the next chapter, a bonus installment on water survival during emergencies, promising further insights to fortify your off-grid water resilience.

BONUS CHAPTER 1 - PREPARING FOR EMERGENCIES

It is often easy to forget that life is full of surprises while we go about our usual daily routines. However, beneath the calm exterior of our lives, there is an undercurrent of uncertainty that can change everything instantly. As Jamais Cascio (2009) said, "Resilience means coping with the unforeseen. Sustainability is about surviving. The aim of resilience is to thrive."

How we deal with unexpected challenges in life reflects how resilient we are. Imagine getting caught in a sudden downpour without an umbrella. It will make you feel uncomfortable, vulnerable, and uncertain. Now, picture yourself in a similar situation in life, and you'll realize how crucial it is to be prepared for the unexpected

Surviving a storm is one thing, but thriving amidst the chaos is a skill honed by those who understand the essence of resilience. When it comes to surviving in rough waters, it's not just about staying afloat. It's about developing the right mindset and acquiring the skills to turn adversity into opportunity..

Preparedness is like a life jacket that keeps us afloat during turbulent times. It is important to be ready for unexpected calamities and natural disasters, especially when it comes to utilizing water resources for survival. By acknowledging the significance of preparedness, we can not only survive but also thrive in the face of adversity.

This bonus chapter delves deeper into systemic preparedness, offering practical strategies for water survival, and demonstrating how a well-prepared mindset can turn unexpected situations into opportunities for growth.

Water Security in Natural Disasters

When nature unleashes unpredictable forces like earthquakes, hurricanes, or floods, securing a reliable water source becomes critical to survival. Knowing how to locate, collect, and store water in these challenging times can mean the difference between resilience and vulnerability. This section aims to empower you with the knowledge and tools needed to stay hydrated amid uncertainty, from identifying water sources to innovative solutions like portable filters and rainwater harvesting.

Locate Water Sources

In the aftermath of a natural disaster, identifying reliable water sources is the first step to survival. Following an earthquake, scout for nearby water bodies like rivers, lakes, or swimming pools, and be aware of any emergency distribution points established by authorities. Knowledge of local lakes and rivers is crucial during hurricanes, and keeping an eye out for emergency distribution points can provide a lifeline. In flood-prone areas, seek higher ground where natural water sources may be less affected, and community wells or emergency water points can be viable options.

Emergency Water Storage

Being prepared with stored water is like having a security blanket during uncertain times. Aim for a 3.8L (1 gallon) of water per person per day for at least three days, stored in durable containers made of food-grade plastic. This ensures a reliable, clean water supply even when regular sources are compromised. In the face of natural disasters, having a well-thought-out emergency water storage plan is a cornerstone of your survival strategy.

Rainwater Harvesting

Harvesting rainwater during and after a natural disaster is a sustainable and valuable practice. Set up containers or barrels to collect rainwater, ensuring they are clean and covered to prevent contamination. Rainwater can serve as an additional source, complementing your stored reserves and providing a natural solution to water scarcity during challenging times.

Dual-Purpose Preparedness

Before the disaster strikes, take advantage of freezing water ahead of time. Fill empty containers and freeze them to serve a dual purpose. As they thaw, frozen water bottles become a backup supply, ensuring you have access to clean water. These frozen containers can also act as makeshift ice packs, keeping perishable items cold in case of power outages.

Portable Water Filters

In any natural disaster, portable water filters are essential for ensuring water safety. Compact and easy to carry, these filters are designed to remove impurities from questionable water sources. Whether after an earthquake, hurricane, or flood, having a reliable portable water filter enhances your ability to access clean water on the go, offering peace of mind during uncertain times.

Water Purification Tablets

Compact and convenient, water purification tablets are valuable additions to your emergency toolkit. Effective in eliminating harmful microorganisms, these tablets are beneficial when dealing with questionable water sources. Whether you're facing the subtle aftershocks of an earthquake, a hurricane, or a flood, water purification tablets provide a compact and reliable means of ensuring the safety of your water supply.

Hydration Packs

Equipping yourself with emergency hydration packs or water bladders is beneficial during natural disasters. These packs are designed for quick and convenient access to water, ensuring you stay hydrated while navigating the aftermath. Lightweight and easy to carry, emergency hydration packs become valuable companions, especially when regular water sources are disrupted.

Involve Your Community

Community collaboration is a powerful tool for resilience in natural disasters. Educate your neighbors about the importance of water preparedness and work together to identify communal water sources. Establishing a community plan for sharing resources, information, and support enhances overall preparedness, turning individual efforts into a collective strength during challenging times.

Remember, preparedness is the anchor that steadies us in the face of uncertainty. Each tip mentioned contributes to building a foundation of resilience. Whether earthquakes shaking the ground, hurricanes unleashing their fury, or floods reshaping landscapes, securing clean water is a lifeline.

With this knowledge, consider the strength in unity, education, and involving your community to enhance overall preparedness. During natural disasters, these tips serve as drops of wisdom, fostering confidence and security drop by drop.

Understanding Dehydration

Dehydration is a severe risk during natural disasters that requires our attention. It occurs when the body loses more fluids than it takes in, disrupting the balance necessary for proper functioning. Water is the body's fuel, and dehydration sets in when it's not adequately replenished.

To understand dehydration better, think of your body as a well-oiled machine; water is the oil that keeps everything running smoothly. Your body becomes dehydrated when you don't drink enough water or lose more fluids than you take in. This can happen for various reasons, such as not drinking enough water, sweating excessively, vomiting, or having diarrhea.

It's crucial to watch for signs and symptoms of dehydration, such as dry mouth, fatigue, dizziness, dark urine, and headaches. If you experience any of these symptoms, increasing your water intake and seeking medical attention if necessary is essential.

Signs and Symptoms

- *Thirst*: The most straightforward signal. When you're thirsty, your body is giving you a clear message—it needs more water. Listen to it.

- *Dark urine*: Pay attention to the color of your urine. Dark yellow or amber-colored urine is often a sign of dehydration. In a well-hydrated state, urine is usually pale yellow.

- *Fatigue and dizziness:* Dehydration can zap your energy levels, making you feel tired and sluggish. You might also experience dizziness or light-headedness.

- *Dry mouth and dry skin:* Insufficient water can cause dry mouth and parched skin. If your lips feel dry or your skin lacks its usual elasticity, it's time to reach for a water bottle.

- *Headaches*: Dehydration can trigger headaches and migraines. Before reaching for pain relievers, try drinking water to see if it helps alleviate the discomfort.

- *Muscle cramps*: Inadequate hydration can cause muscles to cramp and spasm. If you find yourself experiencing unexpected muscle cramps, dehydration could be a contributing factor.

- *Sunken eyes:* The eyes can reveal dehydration. If your eyes appear sunken or have dark circles, it may indicate a lack of fluid balance in the body.

- *Decreased urination*: If you're not visiting the bathroom as often as usual and your urine output has decreased, it might be a sign that your body is conserving water due to dehydration.

- *Rapid heartbeat:* Dehydration can affect the circulatory system, increasing heart rate. Consider your water intake if you notice your heart beating faster than usual.

Prevention Is the Best Medicine

The good news is that preventing dehydration is often as simple as staying mindful of your water intake. Drink water regularly throughout the day, especially if you're physically active or in hot weather. Pay attention to the signs your body sends you—thirst is a fantastic indicator that it's time to grab a glass and quench your body's thirst for hydration.

Whether it's caused by a hot day, vigorous exercise, or an illness, here's a guide to help you combat dehydration with easy and effective first-aid solutions.

Drink Water Gradually: Sip, Don't Guzzle

If you suspect dehydration, start by sipping water slowly rather than chugging it all at once. Drinking too much water too quickly can lead to discomfort and might not be as effective in replenishing lost fluids. Small, frequent sips are the way to go.

Oral Rehydration Solutions: The Right Mix

For a more targeted approach, consider using oral rehydration solutions. These examples contain a balanced mix of electrolytes, like sodium and potassium, which can be particularly helpful in situations where dehydration is accompanied by vomiting or diarrhea.

Coconut Water: Nature's Electrolyte Boost

Coconut water is a natural and tasty way to replenish electrolytes. It contains potassium, sodium, and other essential minerals, making it a refreshing choice for rehydration. Plus, it's a good alternative for those who find plain water unappealing.

Sports Drinks: Replenish Electrolytes

Sports drinks can be beneficial, especially if you've been engaging in intense physical activity. They contain electrolytes that help restore the balance of minerals in your body. However, it's essential to be mindful of the sugar content in some sports drinks, so choose wisely.

Water-Rich Foods: Hydrate from Eating

Certain foods have high water content and can contribute to your overall hydration. Fruits like watermelon, oranges, and strawberries and vegetables like cucumber and celery are excellent choices to munch on for a hydration boost.

Rest: Give Your Body a Break

Sometimes, dehydration is a sign that your body needs a break. If possible, find a cool and shaded place to rest. Taking a break allows your body to recover and can prevent further fluid loss.

Avoid Caffeine and Alcohol: Dehydrators in Disguise

Caffeine and alcohol can contribute to dehydration, so it's wise to limit their intake when you're trying to rehydrate. Opt for water or other hydrating options instead.

Monitor Symptoms: When to Seek Help

It's important to monitor the symptoms of dehydration. While mild dehydration can often be treated with simple remedies, severe dehydration requires immediate medical attention if someone experiences persistent dizziness, a rapid heartbeat, or confusion.

Prevention is critical to avoid dehydration. Awareness of your water intake, particularly during hot weather or physical activity, is crucial. If dehydration does occur, these basic first-aid solutions can help you get back on track to proper hydration.

Addressing Severe Dehydration

Severe dehydration requires prompt and specific treatment to restore the body's balance of fluids and electrolytes. When dehydration reaches a critical stage, seeking medical attention is essential. Here's an easy-to-understand guide on the treatments for severe dehydration:

Intravenous (IV) Fluids: Swift Rehydration

In severe cases of dehydration, especially when someone cannot drink fluids or has lost a significant amount of fluids, medical professionals often administer intravenous (IV) fluids. This involves directly infusing a balanced solution containing water, electrolytes, and sometimes glucose into the bloodstream. IV fluids rapidly rehydrate the body and restore its vital balance (Castera & Borhade, 2022).

Hospitalization: Close Monitoring

Severe dehydration may necessitate hospitalization for close monitoring and intensive care. In a hospital setting, healthcare professionals can carefully observe the individual's condition, administer IV fluids as needed, and address any underlying causes of dehydration.

Oral Rehydration Therapy (ORT): Gradual Restoration

If the severity of dehydration allows for oral intake, healthcare providers may initiate oral rehydration therapy. This involves carefully administering a balanced solution of water, salts, and sugars in specific proportions. While less rapid than IV fluids, ORT can be effective in gradually restoring hydration.

Treating the Underlying Cause: A Holistic Approach

Severe dehydration often indicates an underlying issue, such as persistent vomiting, diarrhea, or a medical condition affecting fluid balance. Identifying and treating the root cause is integral to managing severe dehydration. This may involve medications to control vomiting or diarrhea, addressing infections, or managing chronic conditions contributing to fluid loss.

Monitoring Electrolytes: Balancing Act

Electrolytes, such as sodium, potassium, and chloride, play a crucial role in maintaining the body's balance. Severe dehydration can lead to electrolyte imbalances, which can have serious consequences. Healthcare professionals will monitor electrolyte levels and may provide specific treatments to restore balance if necessary.

Reintroduction of Oral Fluids: Transitioning Safely

As the individual begins to recover, healthcare providers may transition from IV fluids to gradually reintroducing oral fluids. This step is taken cautiously, ensuring that the person can tolerate and retain fluids without worsening dehydration.

Ongoing Medical Care: Prevention

After initial treatment, ongoing medical care may be necessary to monitor the individual's hydration status and address any lingering issues. This may involve follow-up appointments, dietary recommendations, or medication adjustments to prevent a recurrence of severe dehydration.

Remember, severe dehydration is a medical emergency requiring professional attention. If you or someone you know is experiencing symptoms such as extreme thirst, rapid heartbeat, confusion, or fainting, seek immediate medical help. Early intervention prevents complications and facilitates a swift and effective recovery.

Interactive Element: Assessing Your Water Survival Readiness

Are you ready to face the challenges that natural disasters bring? Take this self-quiz to evaluate your water survival skills during earthquakes, hurricanes, and floods. This quiz aims to assess your preparedness level and identify areas where you can improve your water survival abilities.

Answer each question to the best of your ability.

Scenario 1: Earthquake and Broken Water Pipes

Situation: A massive earthquake has damaged the city's water supply lines, leaving you without tap water.

Question: What is the best immediate action?

A. Collect rainwater.

B. Use water from your emergency supply.

C. Try to repair the broken pipe.

D. Wait for external help.

Answer: B. Use water from your emergency supply.

Scenario 2: Flood Contaminates Local Water Source

Situation: Heavy flooding has contaminated your local water source with bacteria and chemicals.

Question: How should you make the water safe for drinking?

A. Boil the water.

B. Filter the water with a cloth.

C. Add bleach.

D. Both A and C.

Answer: D. Both A and C.

Scenario 3: Stranded in a Hurricane With Limited Water

Situation: You are stranded in a building during a hurricane with a limited amount of water.

Question: What is the best way to conserve water?

A. Drink only when thirsty.

B. Ration the water evenly over the expected duration of the hurricane.

C. Drink half the usual amount.

D. Save water for cooking only.

Answer: B. Ration the water evenly over the expected duration of the hurricane.

Scenario 4: Water Supply Cut Off Due to Government Crisis

Situation: A government crisis has led to a cut-off of the water supply.

Question: How can you ensure a continuous water supply?

A. Rely on bottled water.

B. Set up a rainwater collection system.

C. Dig a well.

D. All of the above.

Answer: D. All of the above.

Scenario 5: Contaminated Water in an Urban Survival Situation

Situation: In an urban survival situation, you find that the available water is heavily contaminated.

Question: What is the most effective way to purify the water?

A. Use water purification tablets.

B. Boil the water.

C. Expose water to sunlight.

D. Use a portable water filter.

Answer: D. Use a portable water filter.

Scenario 6: Survival in a Desert Environment

Situation: You're stranded in a desert with limited water resources.

Question: How should you prioritize water usage?

A. For drinking only.

B. For cooking and hygiene.

C. Equally for drinking and cooling your body.

D. Mainly for hydration, sparingly for hygiene.

Answer: A. For drinking only.

Scenario 7: Wildfire Threatens Local Water Supply

Situation: A nearby wildfire has threatened your local water supply, making it unsafe.

Question: What is the best alternative water source?

A. Nearby rivers or lakes.

B. Bottled water, stored as an emergency supply.

C. Collecting rainwater.

D. All of the above.

Answer: D. All of the above.

Scenario 8: Broken Water Filtration System

Situation: Your water filtration system breaks down during a crisis.

Question: What is the best alternative method to purify water?

A. Boil the water.

B. Use chemical disinfectants like iodine or bleach.

C. Purchase bottled water.

D. Both A and B.

Answer: D. Both A and B.

Scenario 9: Lost at Sea with Limited Freshwater

Situation: You are lost at sea with a limited supply of freshwater.

Question: How do you maximize your water supply?

A. Drink seawater when necessary.

B. Ration freshwater strictly.

C. Catch rainwater.

D. Both B and C.

Answer: D. Both B and C.

Wrapping it Up...

In this chapter, we have discussed water survival during emergencies like earthquakes, hurricanes, and floods. We have learned to locate, collect, and store water when unexpected events occur. We have explored various methods of water preparedness, from rainwater harvesting to portable filters.

The central message of this chapter is that being prepared for emergencies is crucial for thriving amidst life's unexpected challenges. Whether earthquakes shake the ground, hurricanes unleash their fury, or floods disrupt the familiar landscape, having a plan and the necessary tools can make all the difference.

In the journey of water survival, preparation is like an anchor that keeps you stable, and the tools ahead are like allies that support you. It is essential to equip ourselves with the necessary knowledge and tools to survive and thrive in the ever-changing currents of life's uncertainties.

In the upcoming second bonus chapter, we will explore a variety of water-saving tools and devices, including innovative filtration systems and efficient water storage solutions. These tools will help you navigate uncertain waters with confidence.

Bonus Chapter 2 - Water-Saving Tools

In our pursuit of sustainable living, every drop of water counts. It all starts with a simple flush. Imagine replacing your toilet and creating a ripple effect beyond your bathroom walls. A study commissioned by Plumbing Manufacturers International found that replacing non-efficient toilets with water-efficient ones can save a staggering 170 billion gallons of potable water annually, equivalent to 465 million gallons daily (Plumbing and Mechanical, 2017).

We must realize that water conservation is not just about numbers. It is also about comprehending the significant impacts of water conservation and how minor adjustments can result in major transformations. In the concluding section of this book, we will discover that water-saving technologies go beyond their environmental advantages and become essential for survival in off-grid living scenarios where water is as valuable as gold.

Discover water-efficient solutions for off-grid living. When water is scarce, every drop counts. Make a conscious decision to explore environmental stewardship and find your lifeline.

Plumbing Fixtures to Save Water

Let's explore some innovative solutions to transform how we consume water in our homes and communities. Discover the importance of selecting the right plumbing fixtures for preserving our planet's most valuable resource.

Pressure-Reducing Valves

Pressure-reducing valves regulate and control the water pressure in your home's pipes. They prevent burst pipes and water wastage and protect appliances from damage caused by excessive water pressure..

Recirculating Hot-Water Systems

Recirculating hot water systems are similar to your home's hot water delivery system. They ensure that you never have to wait for hot water again. These systems constantly circulate water through the pipes, creating a loop that keeps hot water ready. When you turn on the tap, hot water will be readily available. No more wasting water while waiting for it to heat up.

Types of Recirculating Hot-Water Systems:

- *Integrated loop:* This type of system features a pump at the water heater. It circulates water through the hot water pipes and back to the heater, creating a loop. It's a simple and cost-effective solution.

- *Dedicated loop:* In a dedicated loop system, a separate pipe connects the furthest hot-water fixture back to the water heater. This dedicated line ensures a more rapid and targeted hot water delivery to specific areas of your home.

Low-Flow Toilets

Low-flow toilets are designed to use significantly less water than traditional toilets. They optimize the flushing mechanism to achieve the same effectiveness with less water. The most apparent benefit is in the name—water savings. Low-flow toilets can use up to 60-80% less water per flush than their older counterparts. This helps you reduce your water bill and contributes to a more sustainable use of this precious resource.

Performance Showerheads

Modern showerheads use innovative technology to deliver an efficient yet satisfying water flow. They offer features like aerating and pulsating sprays, which give you the experience of standing under a waterfall without wasting water.

Many performance showerheads are also designed to be energy-efficient by mixing air with water, creating a vigorous flow without increasing water usage. This saves water and reduces the energy needed to heat it, making your shower routine more eco-friendly.

Water-Saving Faucets

Water-saving faucets are designed to optimize water usage without compromising performance. They typically incorporate features like aerators, which mix air with water to create a steady and efficient flow.

Types of Water-Saving Faucets:

- *Pull-down:* These faucets have a spray head that can be pulled toward the sink, offering flexibility and ease of use, especially when washing dishes or filling pots.

- *Pull-out:* Similar to pull-down faucets, pull-out faucets have a spray head that can be pulled out. This design is handy for reaching different areas of the sink.

- *Single-handle:* These faucets have a single lever for controlling hot and cold water. They are convenient and have a sleek, minimalistic look.

- *Double-handle:* Faucets with separate handles for hot and cold water give you more precise control over temperature, making them a classic choice for many kitchens and bathrooms.

- *Hands-free:* Touchless or motion-sensor faucets use sensors to start and stop water flow, offering a hygienic and convenient option.

Efficient Water Pumps

The domain of water pumps is where innovation meets necessity. Each pump serves specific purposes based on its design and capabilities, ranging from providing water in remote areas to handling various fluids in industrial processes. Let's also familiarize ourselves with the pros and cons of each type of pump.

Submersible Pumps

Submersible pumps are ideal for extracting water from deep wells or boreholes. They are commonly used in residential, agricultural, and industrial settings where the pump needs to be submerged in water.

Pros:

- Submersible pumps are typically installed underwater, making them ideal for situations where space is limited.

- Being submerged helps to dampen noise, resulting in quieter operation.

- They push water rather than pull it, making them more efficient for deep wells.

Cons:

- If maintenance is needed, retrieving a submersible pump from deep wells can be more labor-intensive.

- They might have a higher upfront cost compared to some other pump types.

Centrifugal Pumps

They are widely used in numerous applications, including water supply, irrigation, drainage, and wastewater treatment. Suitable for scenarios where a continuous, moderate flow is required.

Pros:

- Centrifugal pumps are straightforward in design, making them easy to operate.

- Generally, these pumps are cost-effective and widely used for various applications.

Cons:

- There might be better choices for applications requiring high pressure.

- Centrifugal pumps may need priming to start pumping water effectively.

Hand Pumps

Hand pumps are commonly used in rural areas, emergencies, or off-grid locations where a manual pumping mechanism is essential.

Pros:

- Hand pumps are reliable and can be operated manually without relying on external power sources.
- Often more affordable and simpler than motorized pumps.

Cons:

- Require physical effort to operate, which might not be practical for all situations.
- Typically, hand pumps provide a lower water output compared to motorized pumps.

Remember, choosing the right water pump depends on your needs and the application. Each type has its advantages and disadvantages, and understanding these can help you make an informed decision based on efficiency, cost, and the demands of your water pumping requirements.

Water Conservation Technologies

To tackle the urgent issues of water leaks, shortages, and rising utility bills, let's explore a range of innovative solutions that aim to improve the efficiency of our daily water consumption and promote a more sustainable and responsible future.

Toilet Tank Fill Cycle Diverters

These innovative devices ensure that graywater used to fill the toilet tank is wisely repurposed and directed toward other uses—such as filling the toilet tank— instead of going down the drain. This intelligent diversion maximizes efficiency, combats water shortages, and substantially contributes to water conservation efforts, especially in regions with limited water resources.

Outdoor Irrigation Controls and Rain Sensors

These sensors installed in your garden are equipped with advanced technology that enables them to adjust watering schedules according to the current weather conditions. This unique feature ensures that your garden receives the right amount of water. Also, it helps combat water leaks, thus lowering utility bills. With these intelligent irrigation systems, you can make your garden a beacon of water efficiency.

Soil Moisture Sensors

This innovation aims to measure the soil's moisture levels to provide valuable data for precise watering. Soil moisture sensors are crucial in combating water shortages and promoting healthy, vibrant landscapes by ensuring that plants receive just the right amount of water. These sensors are indispensable tools in the journey toward responsible water usage.

Sprinkler Heads

Modern sprinkler heads are changing the way we water our lawns and gardens. They are designed to distribute water efficiently and evenly, preventing wasteful runoff or overwatering. By upgrading to these advanced systems, you can conserve water, save money on your utility bill, and contribute to a more sustainable way of life. With an efficient sprinkler system, you can transform your outdoor watering routine into a model of water-conscious living.

Toilet Leak Prevention Device

Are you tired of unnoticed leaks in your toilet causing water wastage and high utility bills? Toilet leak prevention devices are here to help. These devices proactively detect and prevent leaks in your toilet, combating water leaks and contributing to water conservation. Using these devices, you can save money on your utility bills and do your part to protect the environment.

Leak Detection System

These sophisticated systems detect potential leaks in real time and alert users, providing a crucial early warning against water wastage. By combating water leaks promptly, these systems minimize the risk of water damage and contribute to overall water conservation efforts, ensuring that every drop is valued.

Water Flow Management Device

Water flow management devices can be crucial to plumbing system efficiency and conservation. These devices are strategically designed to regulate water flow and optimize usage throughout the plumbing network. By ensuring water is distributed precisely, these devices can enhance the overall efficiency of your plumbing system, making them an important essential component of water conservation efforts.

Embracing water conservation technologies helps conserve water, prevent leaks, and save on utility bills.

Wrapping it Up...

Throughout this chapter, we have discussed various water conservation technologies that tackle issues such as leaks, water scarcity, and high utility bills. We have highlighted each tool's simplicity, efficiency, and practicality, ranging from shower regulators to soil moisture sensors. It's important to note that these technologies are more than just about modernizing water usage. Still, they also hold immense significance in off-grid water survival scenarios.

Various tools and technologies become necessary when faced with the challenge of surviving without access to a centralized water supply. By taking a comprehensive approach and incorporating these methods into our daily routines, we can meet our water needs and contribute to the larger goal of sustainable living.

Conclusion

As we reach the final pages of this guide, reflect on the journey we've taken. You've gained the knowledge and tools to navigate toward self-sufficiency by learning about sourcing water, purification, long-term storage, and conservation practices. You protect your health by proactively preventing waterborne diseases and employing innovative water conservation practices to bring you closer to sustainable living.

Being prepared gives you the confidence to face unforeseen challenges in emergencies. With that knowledge and those tools, you can ensure a stable water supply even in unexpected situations. Water-saving tools, including gadgets and practical tools, help you achieve water efficiency and lead a water-conscious lifestyle.

Your dedication to learning about off-grid water survival and your commitment to leading a sustainable and self-sufficient lifestyle are admirable. The knowledge you've gained makes you a responsible steward of yourself, your loved ones, and the environment.

Think about its positive effects on your health, well-being, and the environment. Those close to you will feel more secure knowing you are prepared for emergencies and armed with water survival strategies.

The *Off The Grid – Preppers Water Survival Plan* is more than just a book. It's a guide that leads you towards a future where self-reliance and emergency preparedness come together to form a way of life in harmony with nature. I hope your journey towards this future is healthy, hydrated, and filled with wisdom to sustain the most precious resource of all- water.

Use these insights to navigate off-grid water survival. Your commitment to learning and implementing these principles will contribute to a more water-conscious lifestyle. Here's to hydration, health, and being prepared for any water-related challenges that may come your way.

REFERENCES

American Society for Microbiology. (2011). *What is true for E. coli is true for the elephant.* In PubMed. American Society for Microbiology.

Bhandari, J., Thada, P. K., & DeVos, E. (2022, August 10). *Typhoid fever.* StatPearls Publishing.

Bintsis, T. (2017). *Foodborne pathogens.* AIMS Microbiology, 3(3), 529–563.

Blair, K. (n.d.). *13 Best Water Storage Container Ideas for Long-Term Storage.* EZ Prepping.

Cascio, J. (2009, September 28). *The next big thing: Resilience.* Foreign Policy.

Castera, M. R., & Borhade, M. B. (2022, September 5). *Fluid management.* National Library of Medicine.

CDC. (2020, December 1). *Waterborne disease in the United States* CDC.

Dibley, M. (2022). *Chemical analysis of organic compounds in dew water.* Scholar Works.

Evans, Barbara & Bartram, Jamie & Hunter, Paul & Williams, Ashley & Geere, Jo & Majuru, Batsirai & Bates, Laura & Fisher,

Michael & Overbo, Alycia & Schmidt, Wolf-Peter. (2013). *Public Health and Social Benefits of At-house Water Supplies.*

Garcia-Garcia, D. (2022). *Health promotion and hydration: a systematic review about hydration care.* Florence Nightingale Journal of Nursing, 30(3).

Giardia Infection (2018). *Giardiasis - symptoms and causes.* Mayo Clinic.

Giannella, R. A. (1996). *Salmonella.* NCBI.

Juste, I. (2023, January 31). *Aquatic ecosystems - characteristics, types, flora and fauna.* Thedailyeco.com.

Klobucista, C., & Robinson, K. (2023, April 3). *Water stress: A global problem that's getting worse.* Council on Foreign Relations.

Kristanti, R. A., Hadibarata, T., Syafrudin, M., Yılmaz, M., & Abdullah, S. (2022). *Microbiological contaminants in drinking water: current status and challenges.* Water, air, & soil pollution, *233*(8).

Loucks, D. P., & van Beek, E. (2017). *Water resources planning and management: an overview.* Water Resource systems Planning and Management, 1–49.

Marie, C., & Petri Jr., W. A. (2013). *Amoebic dysentery.* BMJ Clinical Evidence, *2013*, 0918.

Marugán, J., Giannakis, S., McGuigan, K. G., & Polo-López, I. (2020). *Solar disinfection as a water treatment technology.* Encyclopedia of the UN Sustainable Development Goals, 1–16.

Most Common (2019, May 23). *7 Most common waterborne diseases (and how to prevent them).* Lifewater International.

Mealeatey, M. (2021, January 20). *Why failing to prepare means preparing to fail.* Cambodianess.

O'Donnell, D. (2021, June 21). *The importance of water reuse and how it works.* Sensorex Liquid Analysis Technology.

Oberle, S. (2020, March 20). *Water: a universal necessity.* Green Forum. The Green Forum.

Ojeda Rodriguez, J. A., & Kahwaji, C. I. (2020). *Vibrio cholerae.* StatPearls Publishing.

Plumbing and Mechanical. (2017, April 20). *Study: Water-efficient toilets save 170 billion gallons of water per year.* Plumbing & Mechanical.

Prupis, N. (2014, July 14). *"There will be no water" by 2040? Researchers urge global energy paradigm shift.* Common Dreams.

Qian, H., Chen, J., & Howard, K. W. F. (2020). *Assessing groundwater pollution and potential remediation processes in a multi-layer aquifer system.* Environmental Pollution, 263, 114669.

UNESCO. (2023, March 22). *Imminent risk of a global water crisis, warns the UN World Water Development Report 2023.* Unesco.org.

UPMC. (2017, February 3). *Why boiling water makes it safe to drink.* UPMC HealthBeat.

US EPA. (2017, November 2). *Drinking water.* EPA.

Water Science School. (2018, November 6). *Groundwater: What is groundwater?* U.S. Geological Survey.

Woodard, J. (2019, April 15). *What is a UV water purifier and how does it work?* FreshWater Systems.

wikiHow. (n.d.). *How to Drill a Well.* Retrieved January 29, 2024, from

Zahmak, M. (2013, July 2). *The importance of hydration*. Media Relations.

www.ingramcontent.com/pod-product-compliance
Lightning Source LLC
Chambersburg PA
CBHW031152020426
42333CB00013B/633